PRAISE FOR *QUIET STRENGTH*

"As someone who has spent years chasing achievement and wrestling with anxiety, I've come to see equanimity not as passivity but as power. This book is a clear and deeply practical expression of a core dharma truth: We can't control what happens, but we can learn to relate to it wisely. *Quiet Strength* is about the kind of balance that doesn't flatten us but frees us."

—Dan Harris, author of *10% Happier*

"*Quiet Strength* is the mindfulness book of the decade—and the only book that focuses on the magic-like quality of equanimity. Margaret Cullen, one of the nation's most deeply respected mindfulness teachers and pioneers, has created a beautiful masterpiece, integrating wisdom and the emerging science of the mind, with remarkable stories, humor, and humility, and sharing a generous path that is true salve for the soul."

—Elissa Epel, *New York Times*–bestselling coauthor of *The Telomere Effect* and author of *The Stress Prescription*

"As a neuroscientist, I was particularly intrigued by Margaret's reporting on the science of equanimity. She weaves together perennial wisdom, compelling stories shared with real vulnerability, and practical tools drawn from her decades of meditation practice and teaching. This book not only offers a clear and insightful exploration of equanimity but it provides a way into it. The result is a guide that feels both timeless and urgently relevant. You'll want to read it slowly and to return to it often."

—Amishi P. Jha, neuroscientist and author of *Peak Mind*

"Weaving decades of personal practice with the wisdom of teachers spanning spiritual traditions, Cullen's *Quiet Strength* adds richness and depth to the exploration of equanimity and makes it accessible to all, regardless of one's background or religious beliefs."

—Sharon Salzberg, author of *Lovingkindness* and *Real Life*

"A user-friendly, multipronged, methods-rich, and humorous approach to being in wise relationship with deep suffering in all its various forms. At this moment on planet Earth, it is essential for caring humans to embody and enact this 'many doors, one room' non-dual wisdom to whatever degree possible."

—Jon Kabat-Zinn, author of *Wherever You Go, There You Are*

"In *Quiet Strength*, Margaret Cullen explores the profound meaning and far-reaching benefits of equanimity—an often overlooked quality so essential to our well-being. True to the spirit of equanimity itself, Margaret draws from a wide range of traditions to illuminate its nuances, challenges, and methods of cultivation. With warmth and wisdom, she offers a path to this openhearted balance of mind—so urgently needed in these polarized and turbulent times. The courage of her inquiry and the depth of her understanding infuse every page of this transformative work. Highly recommended."

—Joseph Goldstein, author of *Mindfulness*

"*Quiet Strength* is both the medium and the message of this wise and eminently readable book. From its beginning premise that living life in all of its poignancy is an attainable goal, the reader is gently carried along by contemporary personal stories, research conclusions, fundamental Buddhist teachings, as well as universal expressions of wisdom to the recognition that compassion, held in place by equanimity, is the only redemptive response to universal suffering. I love it!"

—Sylvia Boorstein, PhD, cofounding teacher of Spirit Rock Meditation Center and author of *Pay Attention, for Goodness' Sakes*

"This extraordinary book on equanimity has been written from deep inside decades of teaching, practice, and life, as well as the minds and laboratories of thought leaders on the emerging science of equanimity. It is a book all of us need to read in order to actualize equanimity in our own lives and navigate the challenges of our world."

—Roshi Joan Jiko Halifax, abbot of the Upaya Zen Center

"*Quiet Strength* arrives at exactly the right moment. Drawing on decades of teaching, research, and deep practice, Margaret Cullen offers a clear, wise, and practical guide to cultivating equanimity—the steadiness we need to meet life's challenges with clarity, compassion, and resilience."

—Jack Kornfield, author of *A Path with Heart*

"'To be even-minded is the greatest virtue,' said the ancient Greek philosopher Heraclitus, and 2,500 years later, Margaret Cullen beautifully updates, expands, and explains this vital virtue and capacity that is so sorely needed in our contemporary world."

—Roger Walsh, MD, PhD, professor at University of California, Irvine, and cohost of the *Deep Transformation* podcast

"Margaret Cullen has produced a truly wonderful guide to the many faces of equanimity. This book takes us on a journey deep into the heart of this vital skill and is informed by both modern research and ancient spirituality and philosophy. She also takes it very personally and explores equanimity from a very vulnerable, humble, and honest place."

—Kelly Boys, author of *The Blind Spot Effect*

"From her years of practice and experience with these traditions, Margaret Cullen has skillfully interwoven ancient wisdom with modern science to provide a captivating guide to working with our minds through the lens of equanimity. Here is a gentle, accessible, and wise book on how we can use conscious awareness to calm the chaos within and make the world a better place for ourselves and others."

—Paul Gilbert, developer of Compassion Focused Therapy and author of *The Compassionate Mind*

"Equanimity is a flavor of love that cultivates peace of mind. The perfect mix and the right book for troubled times."

—Frank Ostaseski, author of *The Five Invitations*

"In this clear and compassionate book, Margaret Cullen explores equanimity as a powerful inner resource for navigating life's challenges. Grounded in science, philosophy, and contemplative wisdom, *Quiet Strength* offers a practical path to emotional balance and clarity. Both accessible and profound, this book is an essential guide for anyone seeking stability, insight, and resilience in uncertain times."

—Kristin Neff, PhD, author of *Self-Compassion* and *Fierce Self-Compassion*

"As contemplative practice invites an increasing appreciation of the mind's endless gyration, we also gain the gift of equanimity. What is so lovely about *Quiet Strength* is its rich evocation of the micro moments, attitudes, and intentions that place equanimity well within our reach. Combining the wisdom of thought leaders and her own personal experiences, Margaret Cullen shows us how to find a point of balance in these polarizing times."

—Zindel Segal, professor at the University of Toronto Scarborough and coauthor of *Better in Every Sense*

"Equanimity is a truly excellent quality. In *Quiet Strength*, Margaret Cullen splendidly describes equanimity as 'vulnerable without weakness' and 'wise without detachment.' She shows us that equanimity isn't about escaping life—it's about meeting all of it with clarity, steadiness, and delight. This is a book written with heart, wisdom, and enough science to please an old engineer like me."

—Chade-Meng Tan, international bestselling author of *Search Inside Yourself* and *Buddhism for All*

"If all my years in public service and politics have shown me anything, it's that we have a deep need to respond to life's ups and downs with evenness and resilience—with equanimity. To realize it on a daily basis, though, requires habits and practices. Margaret Cullen makes a compelling case for the power of equanimity and offers us many pathways—drawn from her personal experience, as well as spiritual traditions, Stoicism, science, indigenous wisdom, and more—to bring it to bear on our lives. You will enjoy the ride."

—Tim Ryan, former US congressman and author of *Healing America*

"In shaky times, Margaret Cullen helps us find an unshakable inner balance. She integrates science, deep wisdom, and many practical tools into a road map and handbook for navigating our turbulent world. Throughout, her kindness shines through. In every page, she's a friendly, good-humored, and encouraging companion helping us to become less stressed and anxious, and more rested in our own calm strength. Truly, this is a masterwork, and it's needed now more than ever."

—Rick Hanson, PhD, author of *Buddha's Brain,*
Hardwiring Happiness, and *Resilient*

"Equanimity is to feelings what mindfulness is to presence—and *Quiet Strength* is the rare book that makes this truth come alive. With wisdom, clarity, and warmth, Cullen invites readers on a transformative journey, revealing equanimity not as dry neutrality but as a wellspring of empowered engagement. In a world that often feels like it's spinning faster than time itself, this book offers a steady anchor: the quiet strength of balance, presence, and inner calm. A must-read for anyone seeking resilience and peace in the midst of change."

—Amit Sood, MD, CEO of the Global Center
for Resiliency and Wellbeing

QUIET STRENGTH

Find Peace, Feel Alive,
and Love Boundlessly Through
the Power of Equanimity

MARGARET CULLEN
FOREWORD BY DANIEL J. SIEGEL

HarperOne
An Imprint of HarperCollinsPublishers

The credits on page 318 constitute a continuation of this copyright page.

Without limiting the exclusive rights of any author, contributor or the publisher of this publication, any unauthorized use of this publication to train generative artificial intelligence (AI) technologies is expressly prohibited. HarperCollins also exercise their rights under Article 4(3) of the Digital Single Market Directive 2019/790 and expressly reserve this publication from the text and data mining exception.

The material on linked sites referenced in this book is the author's own. HarperCollins disclaims all liability that may result from the use of the material contained at those sites. All such material is supplemental and not part of the book. The author reserves the right to close the website in her sole discretion at any time.

QUIET STRENGTH. Copyright © 2026 by Margaret Cullen. Foreword copyright © 2026 by Daniel J. Siegel. All rights reserved. No part of this book may be used or reproduced in any manner whatsoever without written permission except in the case of brief quotations embodied in critical articles and reviews. For information, address HarperCollins Publishers, 195 Broadway, New York, NY 10007. In Europe, HarperCollins Publishers, Macken House, 39/40 Mayor Street Upper, Dublin 1, D01 C9W8, Ireland.

HarperCollins books may be purchased for educational, business, or sales promotional use. For information, please email the Special Markets Department at SPsales@harpercollins.com.

harpercollins.com

FIRST EDITION

Designed by Jason Kayser

Illustration on page 246 courtesy ImagineDigiCreations, "Kabbalah Elements, Tree of Life, Ten Sefirot, Mysticism, Kabbalah, Sacred Geometry, Jewish Kabbalist, Digital File, Ready to Print, Blessing," Etsy, accessed March 25, 2025, www.etsy.com/ca/listing/1726957105/kabbalah-elements-tree-of-life-ten?ga

Library of Congress Cataloging-in-Publication Data has been applied for.

ISBN 978-0-06-341523-2

Printed in the United States of America

26 27 28 29 30 LBC 5 4 3 2 1

To Sofi,
my greatest teacher

Contents

Foreword by Daniel J. Siegel — xiii
Prologue: Following the Pretense of Accident — 1

Part I: Equanimity: The Forgotten Virtue

1. Why Equanimity? — 11
2. The Worldly Winds — 31
3. Mindfulness and Equanimity Play Well Together — 50
4. This Is Your Brain on Equanimity — 70
5. The Psychology of a Balanced Mind — 88

Part II: Doorways into Equanimity

Introduction to Part II — 109

6. Shifting Perspective — 115
7. Weathering Storms — 128
8. Just Like Me — 132
9. How to Love and Care Without Attachment — 146
10. Stepping Stones to Equanimity — 158
11. Bottom-Up Equanimity — 165
12. Uplifting Stories — 179

CONTENTS

13	Breaking the Spell	188
14	The Serenity Prayer	195
15	Taking Refuge, Finding Equanimity	196
	Before We Move On ...	211

Part III: Real-World Equanimity

16	Equanimity in a World on Fire	215
17	Paradox, Polarization, and ... Uncertainty	237
18	Brokenhearted Equanimity	255
19	Connecting the Dots: Integrity and Equanimity	273
	Epilogue: Drop by Drop	290

Acknowledgments	295
List of Interviewees	297
Notes	299
Credits and Permissions	318

Foreword

by Daniel J. Siegel, MD

Imagine a deep dive into the lived experience, practical application, and science of equanimity, a key to creating a life well lived. Exploring one of the central Buddhist practices of the "four immeasurables," loving-kindness, compassion, joy, and equanimity, this book invites you on a journey that can transform your life. While this magnificent exploration is an invitation to focus on the "quiet strength" of equanimity, it also teaches us to cultivate three keys to a meaningful life: to practice love and kindness for the world, understanding the interwoven relational nature of our experiences; to deeply understand and access the power of compassion to sense, make sense of, and respond to suffering inside our inner and relational lives; and to feel the gratitude and exuberance of joy for this privilege of being alive in this wondrous and waiting world. In many ways, equanimity can be seen as the necessary foundation beneath each of these other learnable and foundational skills of well-being.

In the ancient Greek Stoic tradition, as Marcus Aurelius stated, it was taught that "You have the power over your mind—not outside events. Realize this, and you will find strength." This reflection in his

Meditations is an example of a "consilient" finding—a parallel independent discovery—as it resonates with ancient meditation practices of the East. Epictetus offered a similar Stoic notion: "It's not what happens to you, but how you react that matters."

What are these independently derived perspectives on equanimity teaching us from these wisdom traditions, and why do current scientific studies of mental health suggest that this powerful mental capacity is a key to well-being? Why is equanimity a foundation of health? What are the effective ways we have learned from these studies and from ancient traditions about how to cultivate the emotional regulation and resilience key to mental flourishing?

Margaret Cullen is our deeply qualified, caring, and compassionate guide in this empowering focus on equanimity that effectively addresses these questions with illuminating, inspiring, and instructive answers. Our author speaks directly from her professional experience as a well-trained practitioner and educator in the field of mindfulness and clinical practice. And she powerfully offers her own personal journey to invite us, too, to dive deeply into the inner nature of cultivating equanimity in our lives. Rather than summarize what she has in store for us in the pages ahead, in the honor of writing this foreword, what I'd like to offer to you is a brief sense of what rests beneath the practices and principles we'll be exploring.

Equanimity is often described in various psychological perspectives as the mental capacity to have an open, receptive state of awareness that enables us to be present for whatever arises in life—challenging or rewarding experiences, painful or joyful emotions, failures or successes. This even-keeled composure is not a dissociative state of numbness, detachment, or indifference; equanimity is quite the opposite as it allows us to experience a "bring-it-on" mindset that invites all emotions, mem-

FOREWORD

ories, and responses to lived experience into the knowing of a receptive awareness.

Here is a working definition of *the mind* in the field of interpersonal neurobiology in which I live: "An embodied and relational, emergent self-organizing process that regulates the flow of energy and information." With this definition, we can say that a *healthy mind* is one that cultivates the conditions of *optimizing self-organization*.

Moving as the flow of a river, with one bank chaos and the other rigidity, this central flow of optimal self-organization has the qualities of FACES: flexible, adaptive, coherent, energized, and stable. This FACES flow is a useful definition of mental health. When out of health, we move toward the banks of chaos and/or rigidity. When we optimize self-organization, we experience harmony.

We can use the general term *integration* for how we come to that balance. Neural integration is one way of seeing the inner aspect of mental well-being. This is the "embodied" nature of mind. Relational integration reveals how honoring one another's uniqueness and connecting to one another with compassionate communication is the basis for our relational flourishing. This is the relational aspect of mind, and it includes our connection as individuals to other people and to all of nature.

Here is a simple proposal: Equanimity is how the mind enables *integration to flow* in our lives, both inside the body and in the relational worlds. Integration is the basis of well-being, and this is how and why equanimity is a key foundation of mental health.

How could we learn to cultivate the capacity of equanimity if it isn't already available to us? This is what this book is all about: providing practices and stories that illuminate the path forward. For me as a scientist, therapist, and educator, knowing the foundational mechanisms beneath such practical skills is important so that we can honor what Louis Pasteur once said, "Chance favors the prepared mind." It is my hope that these few words here may help you with the unpredictable events of life, inner and outer, so that you are prepared to take in these important practical lessons so beautifully laid out in the chapters awaiting your journey of discovery.

As we move ahead in our human journey, with all the polarization, isolation, and violence, with the destruction of each other and the natural world in which we are a fundamental part, it is the mental skill of equanimity that will give us a path out of these patterns to live in a new way and with resilience. As we face the rapid technological challenges of artificial intelligence and the ever-increasing distractions of the digital world, we need equanimity now more than ever.

With the equanimity you will cultivate in this important guide, you'll be able to light the flame of hope to keep these possibilities in mind as we move forward, together, into the moments of life ahead. We can shape our lives toward the kindness and compassion that are integration made visible. Equanimity is the foundation for creating such a world.

Prologue

Following the Pretense of Accident

Equanimity. We've all heard of it, but what is it exactly? In its basic meaning, it implies being even-tempered amid the ups and downs of life. You will find out, though—as you embark on a journey deep into the heart of this universally prized quality—that equanimity has tremendous depth and dimension. And, as I think you will also see, developing this quality offers tremendous benefit for your well-being—for everyone's well-being—as attested to by the people from many different fields whom you'll meet in *Quiet Strength*.

What Does Equanimity Promise?

Expansiveness and freedom to love life in all its poignancy; to fearlessly let go of self-limiting definitions; and to see in sadness, fear, and anger some of the same signs of being ferociously alive we feel in delight, connection, and awe.

It has been an unexpected joy to take a deep dive for the past three years into equanimity, a quality that is at once simple and profound—a transformative virtue that changes everything.

It seems that equanimity has been hiding in plain sight for millennia. It's there in every spiritual and sacred tradition; it is a secular virtue celebrated by great philosophers and leaders; and it's at the heart of the meditative tradition. And now it has begun to be studied by a number of researchers in neuroscience and psychology. It has been my pleasure to talk to many of these researchers, as well as faith leaders, meditation teachers, and even one politician, enhancing my understanding at each step of the way so I can share it with you.

How I Got Here

I never really planned to teach meditation, let alone write books about it. A series of fortunate events kept presenting themselves. And now late in my so-called career a project that started out as a modest exploration of perspectives on a Buddhist concept has blossomed into something much larger and richer: an ongoing investigation into the virtue (or habit or way of being) we call equanimity, which I found to be both underappreciated and deeply needed—especially by me. What follows, then, is my journey of learning and experiencing more about equanimity, not simply through a Buddhist lens, but through a human lens, including scientific, psychological, religious, and philosophical perspectives from some pretty astounding people and traditions, as well as the thoughts and perspectives of regular folks, since, as I like to say, we are each of us just another bozo on the bus.

It's not surprising that I would have stumbled into this undertaking more than planned it. In my life, I have pivoted from assistant director in Hollywood to psychotherapist to meditation teacher and writer, and it all happened accidentally (on purpose). As a guest on Sharon Salzberg's podcast, I had the opportunity to verify that "follow the pretense of accident" was wise advice Chögyam Trungpa Rinpoche offered when she asked him how to find a meditation teacher in India. The idea of letting "accidents" happen guided Sharon well, and once she passed it on to me, it also became a guiding principle of my life.

During my first internship facilitating support groups for cancer patients at the Wellness Community, the head of the nonprofit, Harold Rosenberg, said to me one day, "Hey, you're a meditator, aren't you? Someone would like to offer a program called Mindfulness-Based Stress Reduction [MBSR] to our participants. We'd like to pay you to take the class and let us know if it's appropriate for our community." It was the early nineties, and this was only the second MBSR program to be taught in Southern California. Two weeks into the eight-week class, I knew I'd found my calling.

After a move to Northern California, I was lucky enough to launch MBSR at Kaiser Permanente in Oakland, and to help revise the teacher's manual for the Northern California district. I will never forget the thrill of including my meditation experience on my résumé for the job interview. Before long, I found myself teaching meditation and devising contemplative curricula for research studies. One of the first of these, a study looking at mental health outcomes for men with HIV, needed to establish the expertise of the trainers to qualify for a National Institutes of Health grant. For this reason, they paid for my application process to become certified as an MBSR teacher through the Center for Mindfulness, an expense I would have never been able to afford otherwise (and quite an arduous process, I might add). Nobody knows for sure, but I think I was about the tenth person to become certified to teach MBSR. There are now more than thirty thousand MBSR graduates in the US alone.

"Accidents" kept happening. Before we got to know each other, Jon Kabat-Zinn recommended me as co-teacher on a seminal research study,

"Cultivating Emotional Balance (CEB)," out of the University of California, San Francisco (UCSF). It was the brainchild of two very brainy men—emotion research pioneer Paul Ekman and Buddhist scholar Alan Wallace—and was the first study the Dalai Lama personally funded. Those three facts alone meant the pressure was on. Nevertheless, out of this challenging enterprise another fortuitous "accident" emerged. CEB brought together emotion training and meditation in a secular eight-week format, much like MBSR. Knowing that I was trained as a teacher in mindfulness-based programs, a savvy philanthropist interested in scaling the program approached me and asked, "How would you like to be paid to create the curriculum of your dreams?"

Gee—tough question.

The answer became Mindfulness-Based Emotional Balance (MBEB), which brought together much of what I learned from Paul Ekman about working with emotions and emotion theory with the key principles of MBSR.

As a result of a number of research studies, this program gained traction and led to my first book, *The Mindfulness-Based Emotional Balance Workbook* (with Gonzalo Brito Pons). Then, a funny thing happened on my way to Wisconsin. I was invited to make a presentation on MBEB to a gathering of the Mind and Life Institute, founded by the Dalai Lama and others. Rushing to make my flight, I was frazzled by the time I reached my seat. The woman in front of me refused to turn her bag sideways in the overhead compartment in order to make room for my carry-on luggage. Aw, c'mon. It's a small thing. I'm not sure what I said to her, but I know it wasn't nice. The memory is eclipsed by the next moment, when I looked up and saw Thupten Jinpa, the Dalai Lama's translator, beaming at me. I weakly smiled back—we'd never met but I knew who he was—and crumpled into my seat.

The next morning we were sitting around a large conference table in the lab of the world's most renowned meditation researcher, Richie Davidson, who asked if I would start us off with a brief meditation. I complied, made my presentation, and Jinpa gestured to me that he'd like to speak with me in private. I braced myself for an upbraiding in the hallway. Something like "I saw you on the plane, and I don't think you're fit to be a meditation teacher."

Instead, he said, "I've been asked to create a secular training on compassion cultivation at Stanford University. Would you be interested in helping me?"

After twenty years of teaching, developing curricula, training teachers, and writing about mainstream mindfulness programs with diverse populations, I had the great fortune to spend the next fifteen years or so doing the same thing with compassion. And then yet another funny thing happened, which brings us to the book you're holding in your hands. I was relistening to a talk on equanimity by one of my first and favorite meditation teachers, Joseph Goldstein, and a light bulb went off over my head. I could do the same thing for equanimity that Jon Kabat-Zinn had done for mindfulness and Jinpa (and others) had done for compassion. I could take this universal virtue, prized across traditions but especially articulated in Buddhism, and make it accessible.

I created a short workshop—two sessions of two hours each—and piloted it through the Compassion Institute. It was gratifying and remarkable to see the depth of impact such a short program could have. It also quickly became apparent that our times were calling for contemplative tools to develop equanimity. And just as I was thinking deeply about all this, an editor called out of the blue to suggest I write a follow-up book on emotional balance. Of course I was honored, but writing is a love/hate affair for me, so I didn't jump at the chance. And then I thought, *But wait, a book on equanimity!*

While the editor didn't show any interest in the equanimity idea, another happy accident led me to a literary agent who not only liked the idea but also had a larger vision than I had dared to dream of. Before long, I was sitting in the middle of an equanimity retreat, where I'd broken the no-phone rule to await the remote possibility of an offer from a publisher. The phone did ring, as you can tell now. I went a little wild with excitement for a would-be writer on equanimity sitting a ten-day silent retreat.

Although I've always felt a half beat behind the forward momentum of this book, it has been clear to me that it was my job (my karma, my *tikkun olam*) to become its scribe, and for over three years I have served at the pleasure of this book. As mentioned, it began with a fairly small focus: shedding light on the elegant and highly articulated understanding of equanimity within Buddhist teachings. But the book had other ideas.

First, it led me to research equanimity through the lens of the Abrahamic religions. Early on I was led to the writing of Tom Block, an iconoclastic autodidact who was excited to share his years of research on the "mystical fraternity" between Judaism and Islam (spoiler alert: equanimity is one of the links). Fascinated and mind blown, I was quickly hooked on a journey of discovery that was part investigative journalism, part detective mystery, and part exploration on the frontier of an emerging phenomenon. The unadulterated joy of discovery continued with each new alert of cutting-edge research or email saying "Sure, I'd be happy to discuss equanimity and [Christian theology, Nanai culture, neuroscience, systems theory, and so on and so on] with you."

One of the challenges and delights of writing this book has been to synthesize these conversations and perspectives and offer to you, the reader, a comprehensive and cohesive picture of equanimity that ties them all together without minimizing or disrespecting their differences. As Rabbi Tirzah Firestone said in our interview, "It's important not to shrink-wrap Judaism in order to fit it neatly into a Buddhist framework." The challenge, then, has been to find a way to order, organize, and streamline so many perspectives on equanimity without losing the thrill of discovery as each new piece of the puzzle took shape. As it turns out, the puzzle of the many facets of equanimity may actually be more of a prism. As you view it from each new angle, you learn more of its shape and its beauty—and its vital importance to our lives.

In Part I, we'll explore what equanimity is through the lens of religion, philosophy, psychology, and mindfulness—and why it matters. You will hear some of my own personal stories as well as perspectives from the many scholars, teachers, and leaders I interviewed.

Part II is dedicated entirely to a wide range of exercises, tools, and practices to help you cultivate equanimity. Many of these chapters feature meditations, and they represent a variety of doorways into the cool waters of equanimity. Feel free to follow along with the exercises, or access audio versions via the QR code below, or simply read through them first and come back whenever the time feels right.

Part III is where rock meets bone. How can equanimity help when the world seems to be on fire? When I am confronted with a moral dilemma? When things don't make sense? When my heart is broken?

Before you leap into these pages, I want to emphasize one more thing that has been a cornerstone of my journey with equanimity. A huge misconception is that equanimity suggests a state of passivity and inaction, not to mention indifference or a lack of passion. Nothing could be further from the truth. Real equanimity means feeling *all* the feels and yet not clinging to any one of them as permanent or self-defining. It is my deepest hope that this book will offer ways to engage fully with the world rather than withdraw from the world, in a way that brings greater sanity and helps create a future we can love.

PART I

Equanimity: The Forgotten Virtue

Chapter 1

WHY EQUANIMITY?

> Equanimity deepens the poignancy of our lives but
> drains our lives of melodrama.
>
> —Matthew Brensilver

We are in the third week of a silent meditation retreat in Yucca Valley near Joshua Tree in the high desert of Southern California. The hard work of settling my tired, anxious, and distracted mind and body is behind me, and mindfulness has become effortless and self-sustaining. I'm walking in the early morning. It's springtime, and the desert is bursting with life. Sights, sounds, smells, and thoughts appear at each sense door with a vibrancy that borders on the psychedelic. Layers of grime, defenses, and preconceptions have been lifted not only from my senses but also from my guarded heart—leaving it vulnerable to the tender beauty of each cottontail and cactus flower that sparkles with life in the arid desert landscape. The spring is so fleeting here, so unexpected, so beautiful.

After spending a week practicing lovingkindness, Sharon Salzberg has taught us equanimity as a balancing practice. Equanimity is one of a set of wholesome and skillful states of mind that can be cultivated with the help of specific practices. One way of developing equanimity is by silently repeating phrases that serve as guideposts and incline the mind toward that skillful state. Here is the phrase we use to cultivate equanimity as we wish others well:

> Your happiness or unhappiness is more a function of your thoughts and actions than of my wishes for you, and yet I will never cease to wish for your happiness.

I've been sitting with this phrase for several days, directing it toward loved ones, strangers, challenging people, and all beings, so it has begun to have a life of its own. As I walk in the desert, the phrase comes into my mind and is quickly followed by a deeply felt sense of my mother. In the next instant, an insight arises that uncorks a volcano of emotions so strong I'm afraid it might blow me to smithereens. For the first time in my life, I experience, directly, that not only am I *not* responsible for my mother's happiness, but also, what's even more amazing, I can let go of this futile enterprise without forsaking my love for her. I suddenly understand they're not mutually exclusive.

I had agreed, sometime deep in my childhood, to an invisible contract with my mother that it was *my* job to save her from the profound depression and mental illness that had haunted her throughout her life. I had fully believed in the false dichotomy that to forsake this sacred duty was to stop loving her. A lifelong contract—signed in invisible blood, between a little girl (at an age so young I can't even remember) and her parent—that committed the girl to be the one thing standing between her mother and the abyss of her depression and mental illness. This little girl—a ray of sunshine, lifeline, and only hope—didn't even know there was a contract until so many years later, out in the desert near Joshua Tree. The desert absorbed a lot of tears that morning, tears so forceful they pushed their way through years of suppression and denial. Until that moment, I hadn't fully understood the weight of what I had been carrying and how it had affected every corner of my life. As we'll see in Part III, my relationship with my mother involved a shared shoplifting addiction, a food disorder, and a few other transgressive behaviors. There was a lot of buried grief and shame.

This retreat was more than forty years ago, yet it is as vivid to me today as the letters that populate the screen in front of me as I type these words. Afterward, of course, as after any altered state of consciousness, the curtains closed and deeply conditioned beliefs and patterns of relating returned. But an imprint, a template, had been left on my heart and my psyche that could never be erased and that would become the foundation for much of my work and ultimately this book, which is my own answer to the question "Why equanimity?"

What Equanimity Is

Equanimity in English is often defined as mental or emotional stability, particularly in the face of tension or strain. Its key features include tranquility and equilibrium.

I've come to understand equanimity a little differently, as the ability to *fully feel the entire range of human experience*. But feeling without reacting. Reactivity clouds our ability to clearly apprehend what is happening or to discern the most skillful response. Equanimity is tenderhearted without sentimentality, vulnerable without weakness.

Equanimity is wise without detachment, humble without diffidence, and surrendering without passivity.

Equanimity can express itself in micro moments of mindfully relating to the pleasant, unpleasant, or neutral feel of every sense experience; in the ability to take in experience without the ego's need to defend and control information that's flowing in and out; and in recognizing the shared dignity and common humanity of all people everywhere, without exception. Equanimity can also manifest as moments of intrepid fortitude in the face of adversity.

Admittedly, *equanimity* can be a somewhat awkward and even pretentious word. Like many highfalutin terms, it comes from Latin roots: *aequus*, meaning "balance," and *animus*, meaning "spirit" or "internal state." A balanced spirit or state.

I got excited when I found the Latin roots of *equanimity* had these two complementary components. If *animus* is what is alive and animates us, this flies in the face of the idea of equanimity as somehow dull, quiet, not lively. I became more intrigued about equanimity. Can you be calm and animated at the same time? As a "highly sensitive person," I had been afraid that equanimity meant somehow blunting strong feelings and staying in the "middle." Never too high or too low. Reflecting on this Latin root *animus*, I found equanimity became both more appealing and more possible for me. Think of a surfer riding a big wave or a rider on a galloping horse. They are both balanced *and* dynamic. They aren't frozen with fear or squandering energy trying to fight the wave or the animal. They're fully present and pliable, with all their senses open to and receiving every minute change in their environment. Similarly, equanimity doesn't involve being less alive or vital, but more so.

Maya Angelou purportedly said, "Seek patience and passion in equal amounts. Patience alone will not build the temple. Passion alone will destroy its walls." Equanimity does not mean forgoing passion or love. On the contrary: It sets a firm yet fluid foundation for these strong feelings.

A Buddhist Perspective

As we'll see, equanimity is a quality or virtue that many different religions and philosophies throughout the world and across the span of history have prized and cultivated. That said, Buddhism seems to have articulated this quality most fully. Even more important, Buddhism teaches practical tools and techniques to cultivate equanimity.

Buddhist philosophy, cosmology, and phenomenology are complex and vast. In the Visuddhimagga, the great treatise on Theravada Buddhism from the fifth century, Bhikkhu Buddhaghosa defines equanimity in ten different ways. In Pāli (the original language of the *suttas*, the first Buddhist texts), the word most often translated as "equanimity" is *upekkhā*, meaning to observe with even-mindedness or nondiscrimination, which can sometimes express itself as impartiality and at others as seeing the big picture. An entire book could be written to explore and tease apart these different expressions of equanimity, requiring a serious Buddhist scholar who understands Pāli and has devoted many years to the study of the whole system.

For our purposes, we will focus primarily on the two most commonly understood aspects of equanimity: impartiality and nonreactivity. These two dimensions are easy to grasp, easier to cultivate than one might think, and don't require deep Buddhist scholarship. In the second of the three parts of this book, we'll work with a variety of perspectives and practices. I'm confident that at least one of them, and likely more, will click for you and open a portal into experiencing equanimity and the possibility of holding the "full catastrophe" of life as it is—with tenderness and balance, without forsaking love, passion, and the full range of human experience.

Buddhism is also known for its many lists, and equanimity figures prominently on at least four of these, where it is often expressed as the highest virtue or attainment. Of the schemas that include equanimity, the one that most fully articulates it is the four *brahmavihārās* ("celestial abodes," to indicate just how sublime these virtues are):

lovingkindness (*mettā*)
compassion (*karuṇā*)
sympathetic joy (*muditā*)
equanimity (*upekkhā*)

They are considered sublime because their reach is boundless—and are therefore also called the four immeasurables—and because these virtues are the most wholesome ways of relating to ourselves and others.

Each of the four immeasurables has both a far and a near "enemy." The far enemies are fairly obvious: They are the *opposite* of each of the virtues. For example, the far enemy of lovingkindness is hatred, and the far enemy of compassion is cruelty. The near enemies are both more interesting and subtler. They masquerade as the "real thing" but lead us down a completely different path. For example, the near enemy of compassion is pity. Pity is often mistaken for compassion, and yet it separates us from the person suffering rather than connecting us to them. Pity says "poor you" while I remain "up here" untouched by your pain.

The far enemy of equanimity could express itself as either volatility or reactivity. A key distinction here is that it's possible to feel highly aroused on the inside without reacting with volatility on the outside. Later on in this chapter we will tease apart some near enemies of equanimity and explore in detail what equanimity is *not*. For now, let's revel in what equanimity *is*.

Another way to think about *upekkhā* is *seeing without being caught by what we see*. This doesn't mean *not* seeing nor does it mean trying to control or suppress what we *do* see. It's not controlling what we experience but rather relating to it differently—from an open perspective, without grasping after, rejecting, or ignoring it.

Relating to ourselves and our world from an open, nonjudgmental perspective is what mindfulness is all about. More of the world at large has now embraced the practice of mindfulness because of its potential to bring ease into our daily lives. Mindfulness is about creating a space that enables us to be with experience just as it is. As we remove friction in the system by letting go of resistance, our experience naturally changes. It becomes more equanimous. As we'll soon see, mindfulness and equanimity are intimately connected. Sharon Salzberg calls equanimity "the secret ingredient of mindfulness." Mindfulness is both a route to and a function of equanimity. By the same token, equanimity can be a route to and a function of mindfulness. They mutually support and lead to each other.

Concepts like equanimity are often best evoked through metaphors. In Buddhist teachings, a metaphor that is often used for equanimity is grandmotherly love. Whether it is seeing with a grandmother's eyes, or relating through a grandmother's heart (*robai-shin* from Zen Buddhism), the grandmother is able to love fully without being caught in the dramas of her grandchildren's lives. She doesn't take things personally and sees all children as deserving of love and care. She has seen through the trappings of identity and power and has the wisdom of perspective. A grandmother's eyes are quiet eyes, a metaphor that will be picked up by Christian theologian Howard Thurman in chapter 16. They neither seek to possess nor disown. The grandmother's gaze epitomizes unconditional goodwill.

Equanimity's Many Facets

Although there is agreement on the fundamentals of equanimity, different religions and philosophies—even different schools of Buddhism—tend to emphasize distinct facets of equanimity. These distinctions make sense in the context of each tradition and their respective aims, cosmologies, and histories. For example, the equanimity instruction I received highlighted insight into cause and effect (or karma). My story in Yucca Valley is an example of finding equanimity through a liberating insight into cause and effect with my mother.

Vajrayana Buddhism, practiced most widely in the Tibetan tradition, evolved from Mahayana Buddhism, which emphasizes collective awakening and compassion through the ideal of the bodhisattvas, whose compassion is so profound that they "delay" their own enlightenment until all beings are enlightened and commit to liberating every sort of person. It is logical, then, that impartiality (appreciating the worthiness of all beings) would be the aspect of equanimity highlighted in most Mahayana schools of Buddhism, and by extension in Vajrayana. Guided meditations that cultivate kindness, sympathetic joy, and compassion—not just for loved ones but also for strangers, challenging people, and all beings—are all equanimity practices.

Taking another perspective, American Buddhist teacher Shinzen Young defines equanimity as "a radical non-interference with the natural flow of sensory experience." Pause for a moment to consider that. We're usually trying to influence our experience: to get what we want and avoid what we don't want. Non-interference is a radically different approach that is also similar in attitude to impartiality, non-preference, and non-attachment to outcome. It's important to recognize that non-attachment to outcome doesn't mean nonaction or inactivity, as it has often been misinterpreted. We act generally with a result or outcome in mind. But there are many factors aside from our wishes and actions that influence the *actual* outcome. From this perspective, equanimity involves letting go of attachment to an outcome that we don't ultimately control.

Cultivating these qualities isn't as tall an order as it seems. Doing so often involves shifting from a narrower to a broader perspective, which can happen in an instant. On this point, the late Zen teacher Yvonne Rand said something particularly poignant and universal:

Often what is most painful is to reduce ourselves to the one thing that distresses us.

Suffering naturally follows when we narrow our perspective and define ourselves by the one thing that is wrong, allowing this to eclipse everything else. Our narrowed perspective keeps equanimity out of reach.

Just as it is characterized by non-interference and impartiality, equanimity encompasses a larger and wider perspective: taking the proverbial "long view." Putting things in perspective often brings an immediate sense of balance and spaciousness. Though this bigger perspective may not endure—and our moment-to-moment "real life" experience is full of micro moments of both contraction and expansion—having a direct, visceral experience of the difference between a narrow view and a bigger view can serve as a touchstone for what's possible and become part of an internal GPS system for navigating back to equanimity.

What Equanimity Is Not

It's as important to understand what equanimity is *not* as to understand what it *is*. This is especially true because equanimity's subtleties make it subject to frequent misinterpretation.

Equanimity is not being "blissed out" or detached from the world. It is not withdrawing into ourselves and losing touch with the people or issues we care about. In fact, it is most decidedly *not* about "not caring." We may find, as we progress through practices to cultivate equanimity, that it becomes *more possible* to engage with the challenges we face and not to be thrown into chaos or become overwhelmed—to maintain a sense of space and peace. It's never the point of equanimity to become a robot or a statue but rather to become ever more human in the best sense of the word.

Balance is also an essential component of equanimity. Being wobbly is perhaps the most obvious far enemy of balance. However, stasis or stagnation can also be considered a far enemy of balance. Think of our previous image of a surfer riding a wave. A surfer's momentum is critical to the balance that keeps them from sinking. This is, in fact, the opposite of stagnation or stasis.

When talking about balance, a lot of us have an image of the balanced scales of justice, held by a woman in ancient Roman robes who personifies the law. Those scales are static, not moving. We may think that if we can assume that static pose of the woman holding the scales perfectly evenly, that will be equanimity. But people aren't made of stone or marble. We'd actually get very tired holding up those scales.

Shinzen Young sees coagulation as the opposite or far enemy of equanimity, which, as we've seen, he defines as non-interference with the flow of sensory experience. Our experience of ourselves and the world around us is constantly changing. Attempting to stop that flow—interference—is an illuminating way to look at and understand what equanimity is not.

An even more illuminating exercise in learning what equanimity is *not* is looking at its near enemies, states often mistakenly seen as equanimity. Near enemies are deceiving counterfeits. They lead to separation rather than connection and reify the sense of a separate self, all the while innocently masquerading as the "real thing." The most significant near enemies of equanimity are indifference, detachment, apathy, and passivity.

Identifying near enemies clearly reveals and exposes the ways we can get off track with equanimity. Examining them is particularly important because equanimity can be subtle, quiet, soft, and a bit amorphous. Teasing apart these misleading versions of equanimity can lead to what my mentor, the prominent mindfulness teacher Jon Kabat-Zinn, calls an "orthogonal rotation of consciousness." What this refers to is the process of rotating your perspective ever so slightly—leading to a shift in viewpoint that changes everything. Having taught mindfulness, forgiveness, compassion, and equanimity for the past thirty years, I have seen both the stubborn persistence of misconceptions and the power of clear seeing.

A central misunderstanding in mindfulness is that "successful" meditation involves getting rid of thoughts. No matter how much I emphasize, both in instructions and in our discussions before and after meditation practice, that we are changing our relationship to thought rather than getting rid of thought, this misunderstanding persists. In fact, students often believe they heard it in the instructions when it wasn't there!

I've noticed something similar when I teach classes on forgiveness. Students understand intellectually that forgiveness doesn't mean condoning or saying what someone did is OK. It is simply the unburdening of the heart from past resentments. They buy in 100 percent intellectually, but when it comes to doing the practice, the resistance and questions that come up have to do with a mismatch between their emotional experience and their intellectual understanding. In other words, I'm not willing to forgive because I'm afraid that means, somehow, that I'm letting the other person off the hook, even though I understand full well that forgiveness doesn't mean that.

There is an analogy to be drawn with equanimity. Equanimity doesn't mean not having feelings. It doesn't mean not having *strong* feelings. It doesn't mean that we have to *limit* our feelings in any way whatsoever. Equanimity is about our relationship to our feelings; it's not about changing them.

Again, this is easy to understand intellectually and yet, in this same pernicious way, how we embody it can be at odds with how we understand it. This is what I want to tease apart and focus on in this book. For those who would like to cultivate equanimity, to enjoy and share its benefits, there seems to be a gap between hearing and understanding the words and operationalizing them. And it is this gap that fascinates me as a teacher because it's part of the tricky mind that says, *Yeah, yeah, yeah, I get it, but then why can't I do it?* So first and foremost, let's address the gap. Let's put it front and center and look closely at it (*Mind the gap!*), and then let's approach it from lots of different angles.

Indifference and *detachment* mean both disconnected *and* not caring, and they are the most common mistaken interpretations of equanimity. Indeed, equanimity is actually an expression of concern that, far from ignoring its object, connects more fully to it. A classic stereotype of indifference masquerading as equanimity is a stoner saying, "Yeah, it's all good, man," while disappearing more and more deeply into a smoky cloud of unawareness.

We've also seen that apathy and passivity are far from being aspects of equanimity. Balance is an active, dynamic process, as is seen in the image of the surfer and also of a spinning top or gyroscope. The gyroscope maintains balance due to the speed of its spinning. It's an apt metaphor for equanimity because of the elegant mechanism that constantly recalibrates in order to remain upright or on course while forces of varying velocity bear down from all directions. The gyroscope doesn't reject strong bouts of wind, nor does it get attached to calm waters. And, just as a gyroscope only functions in the context of setting a trajectory, likewise setting our intentions helps us maintain balance and equanimity in the face of unpredictable and uncontrollable outward conditions.

Interestingly, a second, lesser-known Pāli word for equanimity is *tatramajjhattatā*, which actually means "to stand in the middle of." Here again, this suggests the opposite of detachment. Instead, equanimity requires us to be up close with experience, standing wholly in the midst of the arising and passing of it all. It involves radical permission to feel, and is, in fact, the opposite of suppression. It frees up coagulated energy and allows for a more fluid and balanced response to what we encounter in our daily lives.

Ironically, even the *Merriam-Webster Dictionary* definition of *equanimity* contains some potential near enemies as synonyms. One synonym is the French word *sangfroid*, literally translated as "cold blood," which is often used in English. While the French word more precisely suggests coolness under pressure, the idea of being cold-blooded would suggest that equanimity involves being ice-cold. But, to the contrary, equanimity involves warmth and active engagement.

Another confusing synonym *Merriam-Webster* gives for *equanimity* is "composure." This, unfortunately, implies the controlling of emotional or mental agitation by an effort of will. Composure often involves willfully trying to regulate our emotions. That's the opposite of flow. It's regulation, tightness, and impedance in the system. Though we typically think of emotion regulation as a good thing, we'll see from Daryl Cameron's research in chapter 18 that emotion regulation can result in feeling less compassion for others. Intentionally regulating emotions is a slippery slope toward suppressing them.

If the dictionary itself offers these synonyms of *equanimity*, then it's not surprising how easy it is to get confused or misled and go down the wrong path when seeking equanimity. It's a dynamic quality that has nothing whatsoever to do with apathy.

Why Equanimity Now?

What does equanimity mean to you, personally? Your family may be in crisis, and the planet certainly is. What if you get a cancer diagnosis? (I myself am a cancer survivor.) What does equanimity look like when the world seems to be on fire?

On a panel for a 2023 book launch at the Rubin Museum, climate scientist and Zen priest Kritee Kanko, whom we will hear more from in chapter 16, was asked quite directly, "What can we do about climate change?" Her answer surprised and delighted me. Instead of ticking off a list of behavioral and dietary changes, she urged us to "feel all the feels."

In an interview with the *Yale Climate Connections* newsletter, she elaborated further:

In neurobiology, when trauma is triggered, it sets the person off into fight, flight, freeze, fawn [i.e., people-pleasing], or disassociate mode. The fundamental teaching of Buddhism is that some kind of human suffering and dissatisfaction is baked into human life. We cannot avoid suffering, but the question is, how can we become OK amidst stress and suffering? When we can face the intense emotions of the climate crisis without triggering a neurobiological response, that's when we can become courageous, clear, grounded members of the climate ... movement.

Kritee is describing equanimity, which does not come from suppression or denial. In fact, as meditation teacher Matthew Brensilver wrote in the epigraph that opened this chapter, it increases the poignancy of our lives while draining them of the drama, urgency, and neurobiological activation that impedes an effective response, whether it be to the climate crisis or a short-tempered loved one.

As Kritee indicates, it takes courage to ride the big waves—in the ocean or in our own hearts. Buddhist teacher Michael Carroll describes the feeling of equanimity as vulnerability: the willingness to stand in the middle of life as it is, "feel all the feels," and respond from a place of tenderness, balance, and wisdom.

The existential crisis of climate change is a prime example of why the cultivation of equanimity, whose value is perennial, is particularly important in our time. In an international study conducted by researchers at Bath University in the UK in 2021, a full three-quarters of respondents agreed with the statement "The future is frightening."

The issue is even more poignant for young people. Among respondents aged sixteen to twenty-five, nearly 60 percent stated that they felt worried or extremely worried about climate change. Three-quarters said they thought the future was frightening, and 56 percent said they thought humanity is doomed. Fear and anger are important survival emotions. They prime us to act. But the existential crisis we face is not something we can fight with our fists or flee from. Freezing won't solve it, either. We need the wisdom and creativity that come from a balanced perspective, able to see the big picture and harness every bit of energy we have and apply it toward creative solutions. For this, equanimity is essential.

Always On

There's an understandable tendency to "catastrophize," as psychologists say, our current ills, whether ecological or political. The digital age has brought incredible resources, including tremendous potential to connect and inform us. What's also clear is that our "always on" culture of social media and twenty-four-hour connectivity has a shadow side.

The first iPhone was introduced in 2007, and in the relatively short time since, the smartphone has wrought a huge change in the way we communicate with others and ourselves. This may be the most rapid adoption ever of a new technology. Our digital devices have revolutionized society, not necessarily for the better.

A study reported in *Psychology Today* found that Americans check their mobile devices on average 159 times a day or nine times per waking hour. This constant smartphone use, it is asserted, "satisfies your hungry neurons with the positive feedback of constant connectivity."

The effects of this behavior, which some might call addictive, are far from neutral. In fact, there is a new diagnostic category—nomophobia—the fear of being detached from mobile phones. In a study recently published in the *Journal of Human Behavior in the Social Environment*, the authors found a significant negative correlation between equanimity and nomophobia across 216 young adults in India.

Social media, which more than half the world's population now utilizes, to the tune of four trillion hours per year, is of special concern. In February 2013, while working at Google, Tristan Harris gave a presentation titled "A Call to Minimize Distraction & Respect Users' Attention." He suggested that Google, Apple, and Facebook should "feel an enormous responsibility" to make sure humanity doesn't spend its days buried in a smartphone.

Harris, who later cofounded the Center for Humane Technology, which produced the movie *The Social Dilemma*, also said in a recent interview that social media is programmed to choose the videos that will most activate your nervous system in what he calls "a race to the bottom of the brain stem."

The twenty-four-hour news cycle also contributes to amplifying volatility and reactivity (far enemies of equanimity). Researchers at Yale, analyzing 12.7 million tweets from more than seven thousand Twitter users, concluded that they tend to express more outrage over time. Why? Users find that posts expressing strong negative emotion, particularly outrage, tend to receive more "likes" and "shares."

In other words, social media often serves as an echo chamber for our worst impulses, a fact that's particularly disturbing when it comes to political discourse. In the same interview mentioned above, Harris spoke about the dangers of social media as a news source, which it has become for so many young people. A few days before the interview, Hamas had invaded Israel in a surprise terrorist attack. A young mother shared with Harris that as she scrolled social media for updates, she was shown videos of violence perpetrated on young children. These clips had been computer "curated" for her as a new mother.

Young people appear to suffer some of the most deleterious effects of social media. In *Generations*, psychologist Jean Twenge wrote, "Every indicator of mental health and psychological well-being has become more negative among teens and young adults since 2012.... The trends are stunning in their consistency, breadth, and size." It was in 2012 that for the first time more than half of all Americans owned a cell phone. In an interview on NPR, Twenge emphasized that "it's not just symptoms that rose, but also behaviors, including emergency room visits for self-harm, for suicide attempts and completed suicides." These effects are by no means limited to the young. Twenge asserts that anxiety, depression, and loneliness have all increased across the board since 2010.

In a *New York Times* opinion piece, Twenge and Jonathan Haidt wrote:

> In 2009, Facebook added the like button, Twitter added the retweet button, and, over the next few years, users' feeds became algorithmicized based on "engagement," which mostly meant a post's ability to trigger emotions.
>
> By 2012, as the world now knows, the major platforms had created an outrage machine that made life online far uglier, faster, more polarized and more likely to incite performative shaming.

Writer and influencer Rebecca Solnit posted this recently on Facebook:

> Is social media just a grand Buddhist scheme to teach me that reactiveness is optional? #thankshaters

The upshot of all this is that not only do we live in a time of existential threats to our survival but also the digital culture in which we all swim has amplified their negative effects. Being always on increases our agitation and distress—the opposite of equanimity. Indeed, despair can compromise our ability to act effectively in the face of these seemingly overwhelming challenges.

This is where the quiet strength of equanimity, once cultivated, can begin to assert itself for our own benefit and for that of the planet. It is the medicine the world needs now. The solutions we need, both personally and globally, will not come from either indifference or volatility. We need tools to "feel all the feels"—as painful as some of them may be—and to act with wisdom and compassion.

And science has demonstrated that this kind of genuine, full-feeling equanimity is possible. In a series of studies at the University of Wisconsin, Antoine Lutz and his colleagues demonstrated, using fMRI imaging, that advanced meditators showed increased arousal to painful stimuli but also quicker recovery and less anticipatory anxiety. These studies beautifully demonstrated how equanimity does not blunt emotions but rather allows us to feel them more fully without getting derailed by them. It is an exciting moment in the history of neuroscience and psychological research that more and more effort is being directed toward understanding our capacities by investigating spiritual and contemplative practices. Equanimity is one of the more recent areas to be researched, and we can expect to see more.

The dialogue between Western science and contemplative practice was jump-started by the Mind & Life Institute, which has facilitated dialogues between His Holiness the Dalai Lama and renowned scientists since 1987. This conversation has been bidirectionally enriching—creating increasing relevance and validation to inner wisdom while making science more multidimensional (first person, second person, *and* third person) and more *fun*. It's fun to learn about the brain. It's fun to learn about how our emotions evolved and function.

National politics in the US and many other countries have sunk to unprecedented extremes of divisiveness, reactivity, and incivility. And the response of media, both conventional and social, is to bombard us with 24/7 doses of negativity designed to activate our nervous systems. Many forces are arrayed to throw us off our game.

Now, more than ever, the world needs every precious drop of balance and wisdom we can offer.

Chapter 2

THE WORLDLY WINDS

> Happy is the man who can endure the highest and lowest fortune. He who has endured such vicissitudes with equanimity has deprived misfortune of its power.
>
> **—Seneca the Younger**

My first forays into solving the puzzle of equanimity—and extending my understanding beyond Buddhism, the tradition I was most familiar with—began with interviewing scholars and leaders from Judaism, Christianity, Islam, and Indigenous traditions. I led with two main questions: "What role does equanimity play in your tradition?" and "What role has it played in your personal life?"

The people I spoke with had a number of things in common: their lives were 100 percent committed to their spiritual path; their paths provided a deep sense of refuge (something we will explore in more detail in chapter 15); most had doctoral degrees and had done extensive scholarship in their tradition; they all respected other religious traditions; they all spoke from a hard-won and deeply personal relationship with their sense of a power and presence greater than themselves.

Over the course of thirty years facilitating support groups for cancer patients, I had numerous occasions to witness firsthand the power of religious faith in supporting people through the most challenging of times. Yet, a part of me remained skeptical, attributing this faith to magical thinking, denial, desperation, or any number of reductive explanations. That is, until I had the opportunity to speak to this remarkable group of people about their personal experience of faith. Without exception, each one spoke from a place so authentic and heartfelt, so well-grounded in scholarship *and* first-person experience, that the only reasonable response was respect and humility in the face of something I couldn't fully understand.

While we may resist the deep truth that our circumstances inevitably alternate—between pleasant and unpleasant, up and down—we know it in our hearts. Buddhist teachings employ a variety of words in a variety of languages for this vital truth. One is the useful (if somewhat old-fashioned) word *vicissitudes*: favorable or unfavorable turns of events, the fluctuating conditions of our lives. A more poetic phrase used in Buddhism is the *worldly winds*, which inevitably sweep into our lives and buffet us around. We all experience these, regardless of race, gender, age, ethnicity, or geography. These vicissitudes are so universal, such a fundamental part of our lives, that they even appear in the form of the classic wedding vow to stay together in both welcome and unwelcome circumstances: "for richer, for poorer, in sickness, and in health."

Part and parcel of the human condition, the worldly winds are hardly restricted to Buddhism. As we'll explore, many different religions and philosophies speak of them and how we can best respond to them.

Buddhist teachings present the worldly winds in four pairs:

pleasure and pain
gain and loss
praise and blame
fame and infamy

Each of us is subject to both sides of the spectrum, the pleasant as well as the unpleasant. As much as we may prefer to experience only what we think of as the *good* rather than the *bad*, we unavoidably encounter both. And indeed, *vicissitude*, which can mean *either* a positive or negative change in fortune, tends to be used more to describe a downturn than an uptick.

It's life's adversities we usually seek refuge from. But it's important to expand our sense of refuge to include *all* the vicissitudes, both positive *and* negative. Equanimity is not only about finding balance in relation to what's difficult. It's also about not getting attached to what's pleasant. This doesn't mean *not* feeling the good stuff. It just means not staying stuck on something that inevitably changes.

There's a frequently retold story about Hakuin, whom some consider to be the most important Zen Buddhist master of the past five hundred years:

> The parents of a village girl barge into his hut and thrust their daughter's newborn child into his hands. They mockingly blame him for fathering the baby. Hakuin makes a deep bow and responds with equanimity, "Is that so?"
>
> Hakuin cares for the child for several years as if she were his own. Eventually, her true father is revealed. The village girl's parents return to Hakuin's hut to reclaim the child, praising him as a great benefactor and asking forgiveness for

besmirching his reputation. As he returns the child, he again responds, "Is that so?"

Hakuin responds to the worldly winds of praise and blame and fame and infamy with perfect equanimity: "Is that so?" It *may* or *may not* be so, but Hakuin, having taken refuge in the Buddhist path, is at ease and at peace with whichever circumstance he faces.

Hakuin can serve as a teacher here, rather than a model. He joined a monastery at age fifteen and dedicated himself to a life of practice and renunciation. Few of us are in a position to choose such a life, but he shows us by example a direction we can follow toward a true refuge from the worldly winds. In the words of the great Thai meditation teacher Ajahn Chah:

> Do everything with a mind that lets go. Don't accept praise or gain or anything else. If you let go a little, you will have a little peace; if you let go a lot, you will have a lot of peace; if you let go completely, you will have complete peace.

Stoicism: Not Good, Not Bad

Stoicism, a Greco-Roman ethical philosophy that's enjoying a recent surge in popularity, places equanimity front and center, advocating emotionally resilient responses to the worldly winds. As the Roman emperor and Stoic philosopher Marcus Aurelius (121–180 CE) put it in his *Meditations*:

> Death and life, success and failure, wealth and poverty, all these happen to good and bad alike, and they are neither noble nor shameful—and hence neither good nor bad.

This list of vicissitudes is remarkably similar to the Buddhist worldly winds. Marcus Aurelius is clearly addressing the existential situation we are all faced with. And, like the Buddhists, the Stoics took refuge in the lawfulness of the natural world as well as in a virtuous relationship to it.

The word *stoic*—as it is most commonly used today—is close to a "near enemy" of equanimity: bearing up with a grim stiff upper lip when the going gets tough. Indeed, the Stoic word for equanimity is *apatheia*, the source of the English word *apathy*. The word for the philosophical school comes from Greek *stoa* and referred to a colonnaded porch, which is where Zeno, the school's first proponent, taught. Being "stoic" in the sense the word means today was not a Greek thing. No one aspired to be like a porch, colonnaded or otherwise.

Putting modern meanings on these words can easily lead to confusion and misinterpretation. The original meaning of the ancient Greek *apatheia* is not indifference but "peace of mind." And a Stoic's stoicism is far more subtle and far-reaching than grinning and bearing it. Indeed, Stoicism is its own version of dharma—the simple truth of the way things are.

In a *Psychology Today* article, "Stoicism as a Fad and a Philosophy," Iskra Fileva says that Stoicism's key promise is "freedom from the anguish and pain associated with the vicissitudes of fortune. These are the rewards of a Stoic sage." Such freedom is what *apatheia* actually means.

Like Buddhism, Stoicism was and is non-theistic. Not *atheistic*; neither tradition makes arguments against the existence of a divine power. Stoicism, in fact, profoundly influenced early Christianity, Judaism, and Sufism, the mystical form of Islam. The New Testament and the works of many of the early Christian writers (themselves composed in Greek) were influenced by ancient Greek philosophy.

During later medieval times, ancient Greek texts were preserved primarily by Islamic scholars and were only rediscovered in the West during the early Italian Renaissance. I've come to appreciate that more fluidity existed among the traditions we've inherited than has often been recognized. If we're going to truly understand—and ultimately embody—equanimity, being trapped in the polarizing dogmatism of a given tradition is unlikely to help. Seeing connections may be more fruitful.

Mystical Connections: Sufism and Judaism

One of the most remarkable discoveries I made while writing this book was of the profound relationship between mystical Judaism and Sufism. As I began researching the historical roots of equanimity within Judaism, I stumbled on an article citing the great thirteenth-century Jewish philosopher Moses Maimonides (1138–1204 CE) as the first person to introduce the idea of equanimity into Jewish thought. Not only did he encounter the idea through his extensive study of Islam, but he also quite literally "borrowed" a Sufi story to illustrate the concept, a story we will see in chapter 12.

Intrigued, I reached out to the author Tom Block, who has studied and written extensively about the hidden and often suppressed connection between mystical Judaism and Sufism, dating all the way back to the inception of Islam in the seventh century. Tom shared his own remarkable story with me about going to Spain to paint in the early 1990s and embarking on a self-guided study of psychology and physics that eventually led him to study various mystical paths, including Sufism, which was totally unknown to him at the time.

As a Jewish man, he was fascinated to discover the historical fraternity between mystics and philosophers from both traditions and how they influenced each other. Fueled by the potential he saw to bring peace and understanding to the seemingly intractable conflict in the Middle East, he began twelve years of study that would eventually lead him to two years in the reading rooms of the Library of Congress, "digging up articles from obscure nineteenth-century journals."

In medieval Islamic Spain, the influence of Sufism spread into Judaism. The two religious traditions were tightly intertwined. In fact, Maimonides, born in Cordoba in Moorish Spain, spoke Arabic before he spoke Hebrew. As Tom wrote in an October 2011 article in *Sophia: The Journal of Traditional Studies*, "Due in part to Maimonides' acceptance of the Islamic ideal, the concept of equanimity (*hishtavut* for Jews) became central to Jewish mysticism, as well."

Although the idea of equanimity was picked up again five hundred years later by the Baal Shem Tov, the eighteenth-century Jewish mystic credited as the founder of Hasidic Judaism, the link with Islam was either repressed or forgotten. In describing *hishtavut* he wrote, "One should think, 'I am speaking only before the Creator, may He be blessed, in order to gratify Him. I am not performing for my fellow men—for what difference does their praise or blame make to me?'"

In our conversation, Tom said to me: "The well that Sufism and therefore Jewish mysticism drew from was the Stoics. I think there was a direct line from Stoicism to Sufism and then Jewish mysticism. The influence of Stoicism is shown in this whole idea of equanimity, humility, and acceptance. Wish for everything to be exactly as it is, and your life will be serene."

Indeed, Maimonides would talk about this as "joyous equanimity in the face of the vicissitudes of life," a sentiment that would be at home in Buddhism and Stoicism alike.

Kabbalah and Mystical Judaism: Devotion

A well-known rabbi who believes himself ready to be initiated into Kabbalah, Jewish mysticism, seeks out and finds a teacher.

> The rabbi says to the teacher, "I'm ready to be initiated into Kabbalah."
>
> "Wonderful. Tell me a little more about yourself. I'm sure within your community you receive a good deal of praise. How is that for you?"
>
> "Well, I'm quite sophisticated at what I do. I know how to work with and counsel people."
>
> "Do you ever get insulted? How is that for you?"
>
> "Being the leader of a community, of course I get insulted. I hate it. It's like arrows being shot into my heart. I can't stand it."
>
> The teacher nods knowingly and says, "It's time for you to go home and work on this for another few years. Come back to me when the praise that you receive and the insults that you receive are just the same and equal unto each other, that there is no ripple."

This story was told to me by Tirzah Firestone, Jungian analyst, author, and founding rabbi of Congregation Nevei Kodesh in Boulder,

Colorado. In the same conversation, she openly shared stories from her own life in which her equanimity was tested. Having lost one child and a brother to suicide, she is no stranger to the vicissitudes of life. And as a rabbi, teacher, writer, activist, and therapist, she has turned back again and again on her relationship with the Divine to stay true to her life's work of relieving suffering.

Although the story she shared has clear similarities to Buddhist and Stoic views of praise, blame, and equanimity, she warned against drawing too strict a parallel. Judaism is a theistic religion, unlike Buddhism. "We're talking about attachment to G-d. It's both transcendent and also extremely personal for people who have developed a constant connection and relationship to the Holy One. For some, the Holy Presence rests upon and within you, keeping you in peace."

As mentioned, the Hebrew word for equanimity is *hishtavut*. The sixteenth-century Kabbalist Rabbi Chaim Vital said: "Behold, after a person is worthy of the secret of *deveikut* [bonding with G-d] one may become worthy of the secret of *hishtavut* [equanimity]."

Equanimity, then, comes about *after* spiritual practice focused on developing a connection with the Divine Presence. Presumably, *deveikut* is what the rabbi in the story at the beginning of this section is told to do before he can be trained in Kabbalah. The equanimity that develops then serves as the basis for further steps along the path, opening channels for wisdom as well as for prophetic understanding.

Hishtavut is a key stage in the development of the Jewish mystical path, which can be measured by treating others equally, whether they've honored or insulted you. Equanimity is also seen as a goal in itself. As the contemporary Kabbalist Moshe Gersht says, "A spiritually classic Jewish concept and character trait, a sense of equanimity, a.k.a. *hishtavut*, is what enables the Jewish people to weather emotional storms with an ever-lowering sense of drama and an always-increasing sense of serenity." (My Jewish family could have benefitted from more *hishtavut*.)

The *deveikut* that leads to *hishtavut* involves prayerful communication with the great mystery. Rabbi Tirzah asserts,

> Making the great mystery personal is not a concept. It's really personal in my life. There's no such thing as equanimity out of the context of the world, which is constantly throwing stones in the lake. And how do we regain our balance? How do we regain that clear surface into which a stone has been thrown? You don't arrive at equanimity. It's a constant dance, a constant arriving.

Sufism: Surrender

Dr. Habīb Todd Boerger, director of spiritual practice with the Center for Spirituality and Practice in Claremont, California, was not a likely candidate to become a Sufi teacher. He was raised in an "exclusivist, conservative Christian" church in a small town in Texas. Like Rabbi Tirzah, Habīb's relationship with God is personal and loving:

> First and foremost, God is merciful, loving, compassionate, gracious, and the source of peace. Equanimity and peace come from placing our heart in a place of safety. And that place of safety is in the womb in which God created us, the womb in which all of creation exists, the womb in which we are all sustained.

For a Sufi, the peace of equanimity originates with Allah. Being Muslim, as the Quran defines it, means "someone who has surrendered to the one God through God's revelation." In other words, "in Islamic Sufism, equanimity comes from being Muslim" in this sense of surrender or devotion to God. The parallels to Kabbalah are unmistakable here, bearing out the assertion of Habīb's spiritual teacher, who told him that "A Jew or a Christian or a Muslim who knows their religion well, knows that there's only one religion, and that's the religion of peace and justice and love and mercy for all without separation."

Allah, while fundamentally the source of peace, has different aspects that bear directly on the practice of equanimity. Habīb referred me to the following passage from Islamic scholar Mumtaz Ali Tajddin:

> The Divine Qualities can be divided into two groups, *jalal* (majesty) and *jamal* (beauty). Majesty, the revelation of which burns and consumes the worlds, is in one aspect rigorous, severe. Beauty on the other hand is the synthesis of mercy, generosity, compassion and all analogous qualities. God has a *jalal* side and *jamal* side, the aspects of Powerful Majesty and Wonderful Kindness, and that these two fall together in Him as *kamal* or perfection. *Jalal* is a masculine aspect, the Overpowering. *Jamal* is a feminine aspect, the loving, kind and beautiful.
>
> In other words, *jamal* is reminiscent of one half of the vicissitudes—pleasure, gain, praise, and fame—while *jalal* loosely corresponds to the other half: pain, loss, blame, and ill repute. Both aspects derive from Allah's underlying peace, justice, love, and mercy. Devotion to Allah

brings about the peace or equanimity needed to experience both aspects and to see them both as divine. As Habīb put it:

> Regardless of what we encounter or experience in our lives, everything is encompassed in that mercy. Everything is encompassed in that womb of love and compassion, and it serves a purpose. It serves the purpose of returning us to our true selves, of purifying the veils that are over our innate primordial nature of goodness, of oneness, of love, of light.

Christianity: Peace

In exploring the Christian view of equanimity, I spoke to two faculty members at the Claremont Theological Seminary, in Los Angeles: Andrew Dreitcer, who is also codirector of its Center for Engaged Compassion, and Aizaiah Yong. Both emphasized the importance Christianity places on peace and its equivalency to equanimity.

My path has crossed on numerous happy occasions with Andrew's over the years due to our shared interest in compassion. Andrew quoted a key passage from the Gospel of John 14:27: "Peace I leave with you; my peace I give to you. I do not give to you as the world gives. Do not let your hearts be troubled, and do not let them be afraid." The peace here is what we know as *shalom* in Hebrew and *salaam* in Arabic, the language of the Quran.

Some of the most famous images in Christianity refer to this peace. Christ himself, in the Sermon on the Mount, said, "Blessed are the peacemakers," referring to the work of actively bringing peace into the world. And while there is in fact no biblical verse that mentions the lion lying down with the lamb, it's a powerful image utilized throughout Christian congregations.

The peace of God is a divine gift freely given. It is not something earned for which we deserve praise. Some Protestant traditions, such as that established by John Calvin, highlight the importance of gratitude for this divine gift as our primary offering to God. In rendering this gratitude, we're neither exaggerating nor diminishing who we are or our own part in this process. Rather, as Andrew put it, we are exercising "clear-eyed discernment, neither inflation on the one hand, nor diminishment on the other, which are flip sides of the same coin." This clear-eyed discernment, he asserts, "is very close to what I understand as equanimity."

Once again, however, equanimity is not a cold state of indifference. Just the opposite, in fact. Andrew stresses that the endpoint or culmination of the deepest contemplation is a warming of the heart, the experience of the warmth of God's presence. When the apostle Paul says (in Phil 4:7), "the peace . . . which surpasses all understanding," it is not simply a cognitive, conceptual, or cold and detached kind of experience. It's heartfelt affection.

In addition to his faculty position, Aizaiah Yong is an ordained Pentecostal minister in the Christian Church (Disciples of Christ). His mother is Mexican American, and his father is Chinese Malaysian. He is devoted to advocacy for all minority and oppressed groups and the possibility of finding healing and love within and beyond faith communities. Like Rabbi Tirzah, Aizaiah stresses the personal dimension of accessing this peace:

> It's that really intimate and personal relationship with Christ through the presence of Christ's spirit that in the Christian

tradition makes the peace of God accessible and available all the time. As we walk with Christ, this peace is available to us, and there is really no condition we walk through that removes us from the reality of peace.

This speaks to the ability to cultivate this peace, this equanimity, through all the vicissitudes and in all the conditions of life.

Another key New Testament passage on this topic is Paul's letter to the Colossians 3:15, in which he tells the congregation, "Let the peace of Christ rule in your hearts." This has been interpreted to mean that bringing the peace of God into your heart is a choice you must make. You have to *accept* the freely given gift of peace. That's our part of the process.

Aizaiah reinforces the notion that peace is "not something we achieve or acquire. It's something we receive through a posture of openness. My sense of contemplative practice in any tradition is really about a posture of openness to a deeper experience."

Spiritual practices such as meeting in community and participating in the eucharist "remind us of Christ's presence in us and with us." As the theologian and priest Raimon Panikkar, one of Aizaiah's influences, puts it, "Peace is connected to reconciliation, bringing together that which is estranged," through a process of compassionate dialogue. Again, peace and equanimity are communal as well as individual practices.

Christ's and God's very nature and identity consist of this peace. As Aizaiah says, it's "available for all life and is present in all of life." It's the ability to preserve a sense of "okayness" even in the midst of suffering. For those who follow Christ, it becomes the foundation undergirding one's life: something that is *given* to us but does not *belong* to us. It transcends possession. Indeed, receiving this peace is "not like we're gaining anything, but it's like we're going a bit deeper into what we've always known."

Indigenous People: The Power of Stories

In Parts II and III you will hear stories from three Indigenous leaders I interviewed about equanimity. Kiliii Yüyan is an award-winning *National Geographic* photojournalist of Chinese and Nanai heritage who tells the stories of Indigenous communities at the ends of the earth through tender and exquisite photographs. Michael Yellow Bird is the former dean of the Faculty of Social Work at the University of Manitoba and a citizen of the Three Affiliated Tribes (Mandan, Hidatsa, and Arikara). He has brought mindfulness and Indigenous contemplative practices together for what he calls "neurodecolonization" to counteract some of the negative mental and physical health impacts of colonization. Part Blackfoot by origin, Charles Lawrence was baptized by traditional Hopi elders, adopted by elders of Lakota and Coast Salish (Musqueam band), and was inspired by his mentor, Joseph Campbell, to become a medicine man.

When asked about equanimity, all three of these Indigenous leaders drew on stories rather than doctrine, and equanimity seemed to be both implicit and embedded throughout their traditions. However, none of the religious leaders I interviewed were either dogmatic or doctrinaire in their approach. It's easy to be put off by the posturing, proselytizing, divisiveness, hypocrisy, and power grabs we've all seen throughout the history of religion. Yet there is a heart within each of these mighty traditions. And right at the center of that heart is equanimity.

I've found it deeply inspiring to discover that equanimity has so many facets, emerging in unique ways in so many places and peoples. It's not one thing; it is many things. And yet, they do converge.

Before moving on, I want to share a few more evocative passages, from India and China, that are brief and close to my heart.

The Bhagavad-Gita, the iconic Hindu text, says,

> A serene spirit accepts pleasure and pain with an even mind, and is unmoved by either. . . . Realize that pleasure and pain, gain and loss, victory and defeat, are all one and the same.

Chuang Tzu, who, along with Lao Tzu, is the leading ancient exponent of Daoism, tells us:

> The sage is quiet because he is not moved,
> Not because he wills to be quiet.

No book on equanimity would be complete without this classic passage, attributed to Jianzhi Sengcan, the third Chinese Zen patriarch:

> The Great Way is not difficult
> for those who have no preferences.
> When not attached to love or hate,
> all is clear and undisguised.
> Separate by the smallest amount, however,
> and you are as far from it as heaven is from earth.
> If you wish to know the truth,
> then hold to no opinions for or against anything.
> To set up what you like against what you dislike
> is the disease of the mind.

Appreciating the Teachings and Wisdom of Many Traditions... While Also Respecting Their Uniqueness and Integrity

As I was trying to puzzle out how equanimity manifests in different traditions throughout the world and throughout history, I took to heart what Rabbi Tirzah Firestone had said to me about not shrink-wrapping Judaism to fit a Buddhist framework. At the same time, I kept seeing so many parallels in the spiritual and philosophical paths of the people I was learning from. How can I point out similarities without conflating everything into one big spiritual pizza with a dozen toppings or reducing each tradition to a fridge magnet slogan ("Be nice."—The Buddha or "Be like me!"—Jesus).

Once again, I followed the pretense of accident. Indeed, a happy accident came in the form of Bruce Alderman, a poet, visionary, scholar, and polymath who is a pioneer in the emerging field of metatheory. To oversimplify, it's the study of how to fairly compare, contrast, and combine knowledge that emerges from a variety of sources. It's important to me, in the context of equanimity, because of the supreme irony that so many nasty (and even deadly) disputes can occur over competing ideologies that seek to foster virtues like peace, love, and, yes, equanimity.

Bruce introduced me to two concepts that began to shed some light on the both/and dilemma I was grappling with. Warning: These concepts are kind of a mouthful, but I have found them powerful aids to allow me to respectfully draw on many traditions.

The first concept comes from Raimon Panikkar (the *same* philosopher and theologian who inspired Aizaiah Yong), known for his pioneering work in interfaith dialogue. Panikkar used the term *homeomorphic equivalence*, borrowed from mathematics, to describe how, for example, Buddha, God, and Allah may serve a similar function *within* their own cosmologies while still retaining their unique attributes. In fact, it behooves us to remain humble about our limited ability to fully understand the identities of these "ultimate" expressions of divinity from the perspective of the onlooker. What we *can* observe—and what was apparent in my interviews with faith leaders—is how faith in these divine figures is embodied through the lived experience of practitioners.

The second concept (using a phrase that Bruce himself coined) describes how a faith tradition succeeds (or fails) to create conditions that allow adherents to achieve spiritual realization *and* also avoid the tendency to impose a judgmental hierarchy about whose faith tradition ranks the best. He defines a *generative (en)closure* as a "zone of intensity" that provides a container that creates the conditions for transformation to occur. The tradition is *generative*: It brings about transformation. It is also *enclosed*: It has the integrity of its methods and ways of seeing the world, but the membrane is not completely cut off from outside. Therefore, we can comment on and compare faith traditions, while still acknowledging that *we don't know how they work from the inside.*

By contrast, a *degenerative (en)closure* fails to provide the necessary conditions for transformation by creating boundaries that are either too tight or too loose. Degenerative (en)closures can be cultlike and judgmental, doctrinaire and dogmatic—impervious to new information and prone to fragility.

When we walk into the temple of a faith we're unfamiliar with, we can appreciate the container its adherents exist in. We can admire the qualities it seems to generate in them. However, we can respect the fact that we can't fully appreciate and judge what we're seeing from the outside in. The proof of the pudding is in the tasting: Does this religion or tradition help its members to fulfill the promise of its path?

From this perspective, we can appreciate the constraints of various traditions as the very thing that enables practice to unfold and helps us refrain from the tendency to disparage them for being closed. We can also avoid the colonial, arrogant, or ethnocentric trap of assuming that *our* set of presuppositions enacted by *our* enclosure allows us to speak with any authority about anyone else's.

Without fail, the adherents I consulted could speak eloquently and with deep inspiration about the transformative qualities of their own traditions *without* arrogance or superiority toward any other tradition or worldview. To appreciate equanimity as it emerges in many different contexts in this book, I want to emulate that attitude. It is a form of equanimity in itself—one that we desperately need in polarized times.

Chapter 3

MINDFULNESS AND EQUANIMITY PLAY WELL TOGETHER

There are two types of meditators. There are beginners and there are experienced beginners.

—Gil Fronsdal

There's good news and there's more good news, and then there's some more good news. If you've already been practicing mindfulness, you've also been cultivating equanimity. If you've never practiced mindfulness, this chapter is full of short accessible practices. Whether you're just encountering these ideas for the first time or revisiting them through the lens of equanimity, like all of us, you possess a beginner's mind that can always benefit from a refresher!

It turns out that mindfulness and equanimity are intimately and inextricably connected, and in fact, in an article in *Frontiers in Psychology*, Juliane Eberth and her colleagues suggest that equanimity is one of the key effects of mindfulness meditation. When I asked Sharon Salzberg to describe how much they would overlap if they were depicted in a Venn diagram, she said: "I think equanimity is the secret ingredient of mindfulness. I would not call it mindfulness without equanimity. It's something else. So, it's completely together. It's like it's one thing."

That's both a practical understanding and an exquisite example of Sharon's no-nonsense straight talk. In that spirit, rather than debate intricate points of Buddhist philosophy, in this chapter we'll take an exploratory journey through the different dimensions of equanimity that are strengthened and accessed through mindfulness practice.

Steadiness of Mind

One of the first capacities cultivated through mindfulness practice is the ability to direct and sustain attention. This is no small accomplishment. Increasingly nowadays, we are actually training our minds to be distracted and to constantly seek new and better forms of stimulation.

In Buddhist teachings—and the Satipaṭṭhāna Sutta in particular—there are four places where we are instructed to direct mindful attention. These are usually taught in a particular order, beginning with awareness of the *body*; proceeding to awareness of *feeling tones*; then, awareness of *states of mind*; and on to awareness of *dharmas*, mental phenomena that underly and order our relationship with reality. They are called the four foundations of mindfulness. We will touch on the first three in this chapter. These foundations are chosen because you get the most bang for your mindfulness buck by systematically focusing your attention on these particular baskets of phenomena. They may sound technical, but I think you'll see that they're actually organic and intuitive.

It's common to begin mindfulness practice by focusing on our most noticeable bodily sensation: the breath. This is a good idea for a number of reasons. While thoughts lead us into the future and the past, the body is *always* in the present moment. The breath is in the body and it is "known" through physical sensations. As much as we may differ in our personalities, frames of reference, physical experience, and so on, we're all breathing. It's accessible to everyone. And, for most people, the breath is pretty neutral and tends not to trigger strong emotions. (Of course, there are exceptions, such as certain types of trauma or acute or chronic conditions like asthma that directly impact breathing.)

Though we can manipulate the breath, we don't have to do anything special to breathe. It happens by itself. We can also feel the sensation of breathing in different parts of the body. When starting out in meditation, it's often easier to feel the rise and fall of the belly rather than the more subtle sensations formed by the breath at the tip of the nostrils, or on the upper lip, or through the sinuses into the throat. The expansion and contraction of the belly often creates a feeling of pressure against a waistband, as well as a more obvious feeling of movement in the body. So this can become the first object we pay attention to. The belly has the added advantage of being physically farther away from where we perceive thinking to happen (i.e., in the brain). It makes it easier to discern when the attention has left the sensations of the belly and gone somewhere else (typically into thought).

I imagine many of you reading this book have probably already done some breath awareness practice. If so, you've seen how insanely hard it is to keep the mind focused on just one in-breath. And how easily the mind gets bored with the breath and looks around for anything even remotely more interesting (like the great mystery of the spot on the carpet in front of your feet—is it a shadow or a stain? Or the fascinating frayed thread on the sleeve of your shirt that is the key to unlocking the great mystery of how the seam was sewn—was it in China? By children?).

If you're an old hand at meditating, you may find, as I mentioned, that going over the basics never gets old. We are all beginners, perpetually.

Your "job" in mindfulness practice is to bring your attention back to the breath, each time it wanders. At first you might be annoyed at how capricious your mind is, how willful and disobedient, how unruly and incapable of staying present. You might get frustrated, impatient and even feel ashamed, convinced that this is a personal problem and your mind is way worse than everyone else's. Lo and behold, this attitude of self-judgment and aversion only gives rise to more thinking and more distraction. Either through skillful guidance, trial and error, or both, you begin to see that the best way to steady the mind is to accept it just as it is. You begin to see that thinking is what the mind does, not *your* mind, *all* minds. That's just the nature of mind. Once you deproblematize thinking, the mind starts to settle down on its own, not because you beat it into submission, but because you created the conditions for it to do so.

Here is the first way mindfulness and equanimity overlap. The successful strategy for sustaining and directing attention toward the object (the breath) is predicated on an attitude of equanimity. It is equanimity that sees the nature of mind and doesn't make it personal. It is equanimity that makes space for the mind to wander without needless annoyance or impatience. It is equanimity that understands the deeper wisdom of true cultivation. As Suzuki Roshi wrote, "To give your sheep or cow a large, spacious meadow is the way to control him." The best way to tame the mind is to give it freedom. Equanimity allows us to embrace such a paradox. We're willing to just go with it.

The following is a brief breath-awareness practice with a particular focus on an attitude of equanimity. Feel free to follow along with the exercise, access an audio version via the QR code on page 7, or simply read through it first and come back whenever the time feels right.

PRACTICE

See the QR code on page 7 to listen to
the audio guide for this practice.

Begin by finding a posture that supports relaxation, alertness, and stillness. For this short practice, you don't need to assume a special position or really do anything other than perhaps sit up a little straighter and shift your body in any ways that will help you feel both relaxed and awake.

Feeling yourself firmly planted on the earth, allow your hands to rest easily and your heart to be soft. Notice the places of contact that your body makes with the floor, chair, whatever surfaces, hardness, softness you're aware of—feeling the simple sensations of pressure, contact.

Now take three deep diaphragmatic breaths. Inhaling through the nostrils and directing the breath down to the base of the abdomen. Filling up the whole torso with the in-breath as if filling up a vessel with water from the bottom to the top. Then on the exhalation, empty the torso of the breath completely as if emptying the vessel of water.

Repeating this for two more complete cycles of in-breaths and out-breaths.

On your third exhalation releasing the breath to its natural rhythm, allowing the belly to be soft and allowing the breath to flow as quickly, slowly, roughly, or smoothly as it does naturally without manipulation.

You're letting your attention rest on the sensations of the breath in the belly and doing your best to feel this in the body. You may or may not have words to describe the sensations of expansion, contraction, or pressure you feel. That's OK, the important thing is to connect with the physical sensations.

For this practice you can either allow the eyes to gently close or keep them slightly open with a neutral soft gaze down toward the floor. Whatever is more comfortable for you.

When you notice your attention has wandered, see how gentle, how easy, and how patient you can be in escorting awareness back to the breath, back to the sensations in the body as you breathe.

There's no need to fight with the mind, to push thoughts away, to cool an agitated mind, to change the emotional state. Just for these few minutes, recognize what's present. Acknowledge it and gently escort the attention back to the breath.

Allowing your awareness to ride the waves of the breath like a boat riding the waves of the sea.

Each new in-breath or out-breath is an opportunity to begin again.

Remembering that it's the nature of mind to wander, not a mistake or a problem. Nothing you need to fix.

When you become aware that the mind has wandered, just escort it back to being aware of the body breathing.

Was that so hard?

Notice how, after only five minutes of practicing breath awareness, you might feel calmer. Why is this? It turns out that a scattered mind is correlated with an agitated body, and a focused mind is correlated with a calm body. Pāli has a wonderfully onomatopoeic word for the proliferation of thought: *papañca*. The more *papañca*, the more agitation.

In the case of breath-awareness practice, equanimity is both the path and the fruit of the path. In other words, the attitude of equanimity facilitates the steadying of the mind and the steadying of the mind facilitates greater equanimity. It's bidirectional. A win-win, plus maybe another win: your friends and family love it when you're a tad steadier.

Non-Reactivity

Mindfulness practice teaches us how to distinguish between pleasant and unpleasant feeling tones and our reactions to them. The feeling tones and the reactions are usually fused together, creating patterns of reactivity that seem hardwired. Feeling tones come in three flavors: pleasant, unpleasant, or neutral. This response is so basic and primitive that it seems even single-cell organisms move toward favorable conditions and away from unfavorable ones. Paying attention to feeling tone, called *vedanā* in Pāli, is the second foundation of mindfulness.

Paying attention to feeling tone has many benefits. For one thing, it short-circuits the "extra" narrative we have around our likes and dislikes. If you are like me and most of my students over the years, you have a pretty brilliant lawyer built into your prefrontal cortex who spends a fair amount of time and energy making a watertight case for why you are 100 percent justified in loving what you love and hating what you hate. Like any good lawyer, this one is expert in making it all about "the other guy." From your lawyer's point of view, you are always innocent and everything desirable and undesirable is happening "out there." But paying attention to feeling tone has a way of cutting through the commentary and directing us right to the source of this particular form of *papañca*. And this source is actually 100 percent *inside* us. In fact, very few of the positive or negative attributes we impute to other people and objects have any objective existence, no matter how convincing our inner attorneys may be. This begins to be revealed to us as we pay attention to feeling tone.

And you can kiss your addiction to drama goodbye while you're at it. Drama can be fun, distracting, and quite addictive. Why do we love to watch medical and courtroom dramas, police procedurals, and soap operas on TV? For the same reason we get addicted to the dramas and soap operas we write and direct in our own minds. Don't be deceived by how boring feeling tones are by comparison. Much like focusing on the breath, paying attention to feeling tones involves a certain amount of withdrawal from drama addiction that can be hard at first, but the payoff is huge. Drama and equanimity are incompatible. Drama activates your nervous system, it fuels reactivity, and it clouds clear seeing. If you want a peaceful heart, you have to be willing to at least recognize, if not relinquish your addiction to drama.

Drama that we are detached from, such as in a stage play, movie, or TV show, can provide a kind of catharsis, most likely because the emotion is not overly personal and solidified, so I'm not suggesting giving up engagement with stories, just that it really helps to see that the drama we invest in easily activates us in ways we're not noticing.

Personally, I'm pretty addicted to detective shows and psychological thrillers, and I'm easily entranced by gossip about the royals or my favorite movie stars. Remember that the role of equanimity is not to eliminate the juiciness of life. Equanimity allows for the fullest possible range of human experience while remembering the dreamlike nature of experience (as in a movie), not to take things so personally, and that everything passes from the stage.

By recognizing feeling tones *before* they blossom into full-blown soap operas, we find much more room to allow them to come and go, without getting caught in aversion or attraction. They lose their power to seduce and overwhelm us. We have greater freedom to see them arise and pass away and can even smile at the endless parade of likes and dislikes constantly presenting themselves to our sense doors.

So how do we do this?

PRACTICE

Begin again with finding a posture that both supports alertness and ease and also announces your intention to be mindful. Taking three deep breaths at the outset serves to soothe the nervous system and reboot the mind.

Now take two to three minutes just focusing on the breath as the primary object of awareness. This time, experiment with using a quiet mental label of "rising" on the in-breath and "falling" on the out-breath as a way to steady and stabilize the attention on the breath, while giving the discursive mind the "job" of labeling.

This noting technique is very quiet, really like a whisper in the mind. Imagine that 95 percent of your energy and attention is involved in feeling the sensations of the breath in the belly and only 5 percent is used for noting the experience.

For the next few minutes, turn your attention toward the feeling tones that accompany every moment of experience. Is this in-breath pleasant, unpleasant, or neutral?

When you become aware of judging mind, what is the feeling tone of that experience? You might notice that you are judging yourself for having so many thoughts. How does that judgment feel?

Experiment with labeling each moment of experience you're able to "catch" with the simple quiet mental label of "pleasant," "unpleasant," or "neutral." If they go by too quickly to label, this is not a problem. The labeling is just a tool; it isn't the point of the practice.

Without trying to analyze or create any particular set of experiences, see if you can shift your attention to the subtle and rapidly changing stream of feeling tones that often occur below the threshold of ordinary awareness.

See if you can discern any tendencies to cling to the pleasant, zone out on the neutral, or push away the unpleasant. If you wake up in the middle of a pleasant fantasy, is there resistance to letting it go? Is there a pull to continue?

If you notice some discomfort in the body, is there a tendency to pull away?

When the breath gets boring, does the mind seek stimulation? Can you return to the feeling of boredom and notice what feeling tone it has? Is boredom neutral, or is it actually unpleasant?

In the last few minutes of the meditation, releasing any effort to notice feeling tones and returning the attention to the simple sensations of the breath.

Please don't be deceived by how simple this may seem. Truly understanding how feeling tone gives rise to attachment and aversion has the potential to transform your life and pretty much comprises the essence of the Four Noble Truths that are the heart of the Buddhist teachings:

1. Human life includes suffering.

2. At the root of suffering is craving or attachment (to hold on to the good and get rid of the bad).

3. It is possible to completely eradicate this kind of suffering.

4. There is a path to follow to get there.

People often get tripped up by the fact that a lot of human suffering is unavoidable and can't be remedied by meditation. Here it can be helpful to make a distinction between *pain* and *suffering*. The pain of breaking a bone or losing a loved one is unavoidable, but the suffering that is added by aversion and attachment can be 100 percent deconditioned. And as we begin to look deeply into feeling tones and how they give rise to worlds of thoughts and feelings, the power of this truth becomes more and more evident.

Non-reactivity is a hallmark of equanimity. It is neither passive nor inactive. It is the capacity to respond rather than react. It's captured in this passage that is widely used in mindfulness programs. Though it's unclear who first said it, it's been attributed to Viktor Frankl without any supporting evidence. It rings true no matter who said it: "Between stimulus and response there is a space. In that space is our power to choose our response. In our response lies our growth and freedom."

At the root of every unskillful action you've ever taken, a feeling tone went unrecognized and led to a thought, a feeling, another thought, another feeling, and so on, and so on. To notice feeling tones is to find that space between stimulus and response—a moment of discernment and equanimity enabling you to act with more wisdom, compassion, and potency.

Opening the Window

Let's explore feeling tone a little more, by looking at a principle known as "the window of tolerance." A big thank-you to Dr. Dan Siegel—clinical professor of psychiatry at the UCLA School of Medicine and author of many books, including *Aware: The Science and Practice of Presence*—for coining this helpful concept. It asks us to consider how we can navigate the stressors and challenges of everyday living most effectively. The window is a zone of arousal wherein we are neither over-aroused, leading to overwhelm, nor under-aroused, leading to shutdown. When we're in the window, we can manage stress most productively.

The arousal could be emotional, physical, psychological, or spiritual. Not only do our bandwidths for tolerance vary from person to person, but they also vary for each of us from day to day and even hour to hour. Though it's rarely helpful to judge our capacity in relation to others, it can be quite helpful to pay attention to how our own capacities shift over time and to track the behaviors and attitudes that allow us to grow our window of tolerance.

Noticing feeling tones is one way to capture that elusive and fleeting moment between stimulus and response. Increasing our window of tolerance is another way, and they are both cultivated in mindfulness meditation.

Blaise Pascal said, "All of humanity's problems stem from man's inability to sit quietly in a room alone." That might seem like a pretty narrow window of tolerance, but put any ten people in a room by themselves for five minutes, take away their phones, and see what happens.

Mindfulness meditation creates the perfect laboratory conditions for learning how to sit with uncomfortable arousal, whether from unpleasant thoughts, uncomfortable sensations, or challenging emotions. With fairly simple instructions about conducive attitudes, posture, and attention training, most people can quickly learn how to move past the first stirrings of resistance and see what happens next.

The process is best if it's gradual, but life is rarely linear. For this reason, it is often wise to begin by practicing with more manageable discomforts and slowly and patiently expand your window of tolerance. Confidence naturally builds as you learn that the sky won't fall if you don't pursue every impulse and you definitely won't die of boredom. One of my favorite meditation tips was the invitation to go ahead and be the first person ever to actually die of boredom.

If you've done any meditation at all, you know how hard it is to sit still. It's not so much that the body resists stillness. It has more to do with the habit of avoiding discomfort. So much of our movement—physical and mental—is driven by escaping discomfort. Often, the discomfort is barely discernable, and we may not even realize we've moved until we suddenly taste ice cream and have no recollection of opening the freezer in the first place.

The following brief practice will focus on physical sensations, but the same instructions apply for uncomfortable thoughts or emotions.

PRACTICE

Once again, shift your posture just enough to signal the intention to be mindful and present. Find a position that allows you to be relaxed and alert. And take three deep cleansing breaths.

Begin by focusing on the breath for a few minutes. If you like the labeling practice, you can note "rising" on the in-breath and "falling" on the out-breath. If the noting practice isn't helpful in steadying the mind, just feel the body breathing as best you can.

When the mind wanders, don't take it personally. It's just doing what minds do. See if you can escort your attention back to the breath without judgment or impatience, no matter how frequently it gets distracted.

Now, just for a few minutes, set a strong intention not to move, to sit still. Very quickly you are likely to notice some physical discomfort—perhaps an itch, or a signal from your hand or foot that it needs to move slightly to relieve some uncomfortable sensation of pressure. Maybe your eyes are insisting that you open them for a moment, or your back wants you to adjust your posture slightly.

Experiment with resisting the urge to move and allow the uncomfortable sensation to become the object of your attention, knowing it will only be for a few minutes. See if your mind can remain steady as you explore the elements of the sensation: Is it hot, pulsing, pressing, itching?

Here again, you can use the labeling technique if that helps you to remain more neutral toward the experience.

Now notice what happens to it upon observation. When you notice an unpleasant experience with a calm mind, what happens to the experience? Does it get bigger? Does it diminish? Does it change or move to another part of the body? Does it vanish altogether? There is no right or wrong answer here, there is just tracking what is true in your own experience.

If the sensation vanishes, just come back to the breath until another comes along. And it will, that's a promise. When it does, once again turn your attention fully to the experience. Dare yourself to die of itching without scratching.

Once you've had a chance to explore the experience of non-reactivity to discomfort, return your attention to the breath for a moment or two before completing the practice.

This might be a sad statement about my meditation practice, but one of the more powerful revelations I've had has been about not scratching itches. Who knew they actually went away more completely *when you don't* scratch them! Even more important, most of our desires are like itches we feel compelled to scratch. It seems to us that the itching is at the root of our discomfort, but what's actually causing the pain is the desire for something different. It's the resistance and aversion to the present moment, the desire to be or feel or have something other than what we have. Scratching is like licking the ice cream, or buying the sweater, or following whatever impulse might be tempting us. At the moment we get the taste or the thing, the pain stops, but not because we got the thing, because the pain of desire stopped.

It turns out desires go away even when we don't fulfill them. Their nature is to come and go, like everything else. Not only that, when you scratch an itch, it feels better temporarily but tends to itch more, *not* less, after you've scratched it. When you really think about it, this is a truly radical insight that has the potential to dismantle our materialistic culture. Consumerism is built on the false foundation that the discomfort of desire is cured by attaining the object of desire. And, of course, the evil genius here is that because we feel momentary relief from desire when we get the object, we associate the good feeling with getting the thing, rather than with the momentary end of desire, thus seeding the roots of addiction.

When I first began teaching MBSR (Mindfulness-Based Stress Reduction) almost twenty-five years ago, meditation was just becoming a "thing." I was both amused and appalled to see that the advertising business had quickly pivoted from visuals that promised wealth and the good life to images of people meditating. This had to be the ultimate perversion of consumerism. Suddenly beautiful blondes were appearing cross-legged with their eyes closed on the back of shiny new pickup trucks. My very favorite was an ad from a 1999 catalog for Nordstrom, the department store, with a very young dreamy-looking man in an expensive cashmere sweater. He couldn't have been more than nineteen or twenty years old and he was looking out a window at the sea with a wistful expression.

The copy read:

I am inspired by the simple things, the bright white of the sand, the warmth of the sun, the serenity of the waves. I have matured [yeah, right!]. I surround myself with people I love, things I truly need [like the cashmere sweater]. Finally, I am enlightened.

Yeah, and wow. This is verbatim copy from the 1999 print ad. If only a cashmere sweater ever did that for me. Believe me, it wasn't for lack of shopping.

Mindfulness Leads to Wisdom

If you were to take a class or go on retreat at a meditation center like Spirit Rock in California or the Insight Meditation Society in Massachusetts, the kind of meditation you would learn is called *Vipassana*, a Pāli word that is generally translated as *insight*. This type of mindfulness is based on the foundations of mindfulness we have been exploring, and it is specifically designed to generate insights that lead to profound wisdom about the nature of reality, and wisdom is at the very heart of equanimity.

EQUANIMITY IS . . .

Wisdom that allows us to maintain perspective and discern the most skillful response.

Wisdom that sees the larger perspective and keeps us centered when the winds of change blow us hither and yon.

Wisdom that keeps our inner gyroscope pointing us in the right direction.

Wisdom that allows us to let go of the many things we can't control and thread the needle between clear seeing and cynicism.

Wisdom that creates the spaciousness of mind to not take things personally, even our own thoughts!

Wisdom that recognizes that pain is unavoidable but suffering can be greatly reduced, if not altogether vanquished.

So how does paying mindful attention to the foundations of mindfulness lead to wisdom? As we've seen, focusing on the body leads not only to steadiness of mind but also to the possibility of experiencing even our own sensations as phenomena that are as much a part of nature as they are a part of our bodies. And, just like nature, they are constantly changing.

In the second foundation, *vedanā*, feeling tone, we begin to see how dissatisfaction can run our lives and give rise to confusion about the true source of our own happiness. By shifting to a neutral relationship to pleasant and unpleasant experiences, we begin to see the arbitrariness of our likes and dislikes and to sever the deeply conditioned patterns of reactivity that are built on ignorance and confusion.

In the third foundation of mindfulness (*citta*, mind), we direct mindful attention to more subtle phenomena: thoughts, emotions, and mental states. Paying attention to thoughts as objects of awareness can be elusive at first, both because thoughts are intangible and because it can feel like we are using thinking to watch itself. But awareness actually isn't the same thing as thinking. It is like an image with a built-in figure/ground reversal that reveals an entirely different picture once you're able to see it.

This is also where we tend to be the most identified, naturally taking our thoughts personally and believing them to be an accurate reflection of reality. Rumination is a hallmark of depression and anxiety, and it is fueled by identification with our thoughts. When Zindel Segal and colleagues adapted MBSR for people with depression, a key mechanism of the success of Mindfulness-Based Cognitive Therapy (MBCT) for people with depression and anxiety turned out to be the power of "decentering," a psychological term for *not taking our thoughts personally*.

So let's do another brief practice, dipping our toes into the subtler territory of thought that is at the heart of the third foundation of mindfulness.

PRACTICE

Begin as usual with a shift in posture and three deep diaphragmatic breaths and then spend a few minutes just paying attention to the sensations of the breath in the body.

When you're ready, apply a little figure/ground reversal so that, when you notice a thought has pulled your attention away from the breath, stay with the thought, allowing it to replace the breath as your primary focus, without trying to get rid of it and without allowing yourself to get carried away by the thought.

See what happens to your thoughts upon observation.

It can be helpful to label thoughts as they arise. You might use a general label, such as "thinking," or a more specific one, such as "planning," "judging," or "worrying." For example, the thought *This is really boring* can be labeled as "judging." If choosing specific labels gives rise to more discursive thinking, keep it simple and use the general label "thinking."

Each time a thought arises, give it a label, observing what happens to it without analyzing it or letting it take the mind away. Notice the contour or category of thought, rather than being pulled into the content of the story.

If you find that a thought vanishes upon observation, just return your attention to the breath until another thought comes along.

Thoughts can be very seductive. You may find it helpful to think of attention like a spotlight, which can be turned toward the primary object of meditation, whether that is the breath, sound, sensation, or thoughts. This spotlight allows experience to be seen in the clear light of nonjudgmental awareness.

There is no need to look for thoughts, or to push them away. Using the breath as the primary object of awareness, allow the attention to turn fully toward the thoughts the moment you become aware of thinking. Then, in addition to labeling the thoughts, notice what happens next.

Close the meditation with a minute or two of returning attention to the simple sensations of the breath.

It is hard to overestimate the power of not believing everything we think. Not only does rumination play a critical role in exacerbating anxiety and depression, judging thoughts increase anger, doubting thoughts increase shame and insecurity, and so on.

Maybe twenty years ago, I recall a meditation teacher saying, "I've been meditating for twenty-five years. I don't have any fewer judging thoughts. I just don't believe them anymore."

Like them, I'm an expert judger. At various retreats I've been instructed to just watch judging thoughts, and apply the label "judging, judging, judging" whenever one appears. It sure helps to have a sense of humor when you practice this way. There are so many judgy thoughts and they're so arbitrary. Seeing how empty they are is liberating. Learning not to believe your thoughts is subtly but crucially different from trying to change your thoughts, or get rid of them. You have to use clever child psychology with your thoughts or they'll outsmart you every time. In relation to your thoughts, you are the kind, wise, bemused grandparent, offering unconditional love and goodwill to the toddlers running around unsupervised in your brain.

And speaking of the brain, that's the next place our journey into equanimity takes us: What are researchers from the new field of contemplative neuroscience learning about the equanimous brain? What questions are they asking? What are they recommending?

Chapter 4

THIS IS YOUR BRAIN ON EQUANIMITY

*Although involved in worldly ways,
unshaken the mind remains.
And, beyond all sorrow, spotless, secure.
These are the highest blessings.*

—Mangala Sutta

My quest to learn as much as possible about the quiet strength of equanimity has led me not simply to spiritual teachers and thinkers but also to various kinds of research, including in particular neuroscience and psychology, which is what these next two chapters delve into. In fact, I found there has been little formal research into equanimity per se. A search of the National Institutes of Health's PubMed database, going back to 1940, turns up only forty-one articles with *equanimity* in the title and only eighteen articles published in the past ten years.

One of the things this tells us is that the specific "construct" labeled "equanimity" is not something that in scientific research terms has a widely accepted definition. It's also the case that the specific skill or quality of equanimity is not easily distinguished from mindfulness, which has been more widely studied. As of early 2025, PubMed lists 199 total studies with *equanimity* as the keyword compared to 32,200 for *mindfulness*. When I started sniffing around, though, I found that interest in equanimity as a subject worthy of study is increasing, for the very reason I believe that is starting to attract more people to equanimity itself: It appears to be something we need now, perhaps more than ever.

The field I dip my toe into in this chapter is broadly labeled as "contemplative neuroscience": the study of what is happening in the brain and body when we engage in contemplative practices and traditions, such as meditation and yoga, and therefore also what the human mind is capable of under certain kinds of influences. It is in its infancy, beginning very slowly in the 1970s but gaining a lot of momentum once the Dalai Lama took an interest in it and helped to found the Mind & Life Institute in 1987.

It now encompasses a wide range of practitioners, including, as we will see, some folks who believe in using medical-type devices to aid in the pursuit of contemplative practice. Naturally, there are plenty of disputes and disagreements in an arena that is trying to study something so amorphous. My good friend Cliff Saron, neuroscientist at UC Davis's Center for Mind and Brain, for example, operates with a robust skepticism about claims and labels for certain kinds of experiences. How do you really know what someone has experienced or attained?

While some researchers are attempting to make brain maps of adept meditator brains, others, like Cliff, look to find models that look not only at the whole person but also at the whole person as inextricably bound to context and culture. Real progress in a contemplative discipline like meditation, in their view, is only fully understandable through a big lens that includes a person's life and behavior when they're not engaging in a formal practice. Just as it's dangerous to reduce anyone to a label, such as their disease condition, it's also unsound to treat the words we use to describe mental activity as if they had simple one-dimensional meanings.

As you dive into these chapters, then, please appreciate that *equanimity*, after all, is a word like any other, with a wide range of meanings, and we treat it as overly concrete at our peril. As I will emphasize throughout this book, it's multifaceted and wide-ranging, not one simple thing. Keeping that in mind, let's enjoy the ride.

Zapped with Equanimity?

It's a beautiful day in Tucson, Arizona—cerulean-blue sky, low seventies, gentle breeze. I'm excited and a bit nervous as I leave my hotel extra early and follow the clear directions to neuroscientist Jay Sanguinetti's Sonication-Enhanced Mindfulness Acquisition (SEMA) lab at the University of Arizona. As I exit the free tram on the far side of campus, I look around for the blue building. Behind me is a low-slung complex that looks a bit like a seedy motel. That can't be it. Jay's research coordinator, Erica, meets me outside and escorts me into the modest complex. I enter a small warren of rooms that have the vibe of a very lived-in college dorm.

It's a bit awkward at first. I have boldly and uncharacteristically (the book made me do it!) invited myself to visit the lab so I could experience the transcranial ultrasound they're experimenting with. (*Transcranial* simply means it's applied through the skull.) Jay has been working closely with meditation teacher Shinzen Young to develop a combination of bot-driven meditation instructions and transcranial ultrasound modulation to evoke equanimity. Naturally, I'm intrigued!

Jay has been hard to pin down. He's a busy guy with a university lab to run, a young family, and a Silicon Valley start-up. But Shinzen, whom I've known since his days back in Los Angeles, has been extremely generous with his time and has spoken to me for hours in advance of this trip.

Erica quickly makes herself scarce, so I end up spending the better part of the day with Jay, Shinzen, and Brian (a PhD student), and it feels a bit like visiting the alien planet of *The Big Bang Theory*, but in a really sweet and intellectually stimulating way. There's clearly a lot of brainpower in these shabby rooms.

I've come equipped with some pointed questions, thanks to Cliff. For example, "Is there a phenomenological account of sham-controlled stimulation that is analyzed blind and then rigorously coded by people not involved in data collection?"

Foolishly, I share this as my opening gambit, probably trying to prove I'm smart enough to be in the room with these brainiacs, but it's clearly not time to get into deep questions. It's time to take a test run. After an extemporaneous dharma talk from Shinzen (which everybody takes out their phones to record), Brian and I get down to the business of stimulating my brain.

"Did you happen to bring a brain scan?" he asks, in all seriousness.

"Uh, no, don't happen to have one."

"That's OK, we'll just use average brain measurements."

Pointed questions notwithstanding, I feel perfectly comfortable admitting that my brain is average and that should be just fine.

Brain stimulation has been used to address mental health problems for decades. At its most invasive, electroconvulsive therapy (ECT) sent electric currents through the brain that could relieve severe depression and treat some forms of psychosis. However, the electric current often did some damage along the way, sometimes wiping out swaths of memories. Over the years, ECT has become more targeted—causing less collateral damage—and safe and novel forms of electric brain stimulation are being developed and studied. More recently, one less invasive form of brain stimulation, using magnetism, has been increasingly touted for treatment-resistant depression when pharmaceuticals don't work. There's much less risk of brain damage, but it's a major commitment of time and money if insurance doesn't cover it and the results are not a slam dunk.

Diagnostic ultrasound has a long history of being safe and effective in everything from pregnancy scans to echocardiograms. Less known—but nonetheless around for more than one hundred years—is the use of *interventional* ultrasound to not only scan but to modulate the nervous system. It sends non-destructive yet highly focused acoustic energy into deep brain regions such as the posterior cingulate cortex and the thalamus. In Jay's view, this promises to be perhaps the safest and most accessible transcranial intervention for three reasons: the technology is widely available, the sound waves are non-invasive, and the sound waves hit only the areas that make a difference, avoiding side effects that might come from a less-targeted application.

I remind myself of this as Brian applies a big glob of cold gel to the back of my freshly washed hair (anything for the cause). He clearly and calmly explains what's about to happen, indicating that some people report a change in their relationship to thoughts. They report that the thoughts become less "sticky," which they claim is similar to the experience they've had of being in deep retreat.

I'm eager. We spend about fifteen minutes with one probe gently moving around the back of my head and another one stuck on my forehead. I focus on my breathing and wait. Nothing happens. We're done, and he suggests I go meditate for a bit and leads me into a windowless room with furniture that belongs in a squat. I overcome my squeamishness and settle into about twenty minutes of meditation practice. Still nothing happens.

The partnership between Jay and Shinzen began when Jay attended a two-day retreat with Shinzen in 2014. Since he found the teaching there radically different from anything he'd experienced before, he thought he might have found a kindred spirit. After the retreat he suggested the idea that ultrasounding the brain might help in teaching people meditation. At the time, Shinzen thought it was too early and more needed to be learned. By 2017, they had reconnected. They felt it was worth seeing what would happen if they targeted the posterior cingulate cortex, which is associated with the *default mode network*, active when we're daydreaming or mind-wandering. Could targeted ultrasound quiet the inner chatter there (Buddhism's monkey mind), facilitate meditation experience, and put us in touch with the natural equanimity at the base of our experience?

When it's time for lunch, Shinzen proudly announces that his new home of Tucson has been recognized by UNESCO as a City of Gastronomy (oddly there are only two in the US). We pull into a nondescript strip mall with as much charm as the lab's office building and proceed to have a world-class dim sum lunch, which Shinzen orders in Chinese. He likes to say that he is a "Buddhist-informed mindfulness teacher who got turned on to comparative mysticism by an Irish Catholic priest and who has developed a Burmese and Japanese fusion practice inspired by the spirit of quantified science." Appropriately, then, he has a knowledge base that ranges from languages and linguistics to mathematics, to Buddhism, to systems theory, to comparative religion, to physics, and beyond. He is like having your own personal impish ChatGPT—only more reliably accurate (and more talkative).

We savor the food and geek out on the conversation, which pinballs between meditation, neuroscience, psychedelics, the Dalai

Lama, Buddhist philosophy, mathematics: the usual suspects. I learn the origin story of this ambitious project and listen with fascination and a bit of envy to the experiences they've had using ultrasound on themselves. Maybe my brain wasn't average after all (ha ha) and that's why I didn't feel anything?

In Shinzen's Unified Mindfulness approach there are three main components: concentration power, sensory clarity, and equanimity. In a recent talk at a scientific conference, he said, "If I could choose only one of these three it would be equanimity." Indeed, this is what he's choosing to focus on in his wide-ranging collaborations with researchers at several major universities.

He loves using metaphors from the physical sciences when describing equanimity. They can be a little technical and require some unpacking for anyone who has forgotten high school physics, but they come from a mind that sees a very big picture. To name a few:

- Reducing friction in a mechanical system (Equanimity = $1/F$)
- Reducing viscosity in a hydrodynamic system (Equanimity = $1/\mu$)
- Reducing resistance in a DC circuit (Equanimity = $1/R$)

Extending these metaphors and taking some Shinzenesque big mental leaps, he contends that "perfect equanimity would be analogous to 'superconductivity' within all your sensory circuits." Because equanimity, seen in this way, readily maps on to existing scientific understanding about phenomena like efficiency, flow of information, and the structure of biological networks, he believes mainstream science will zero in on it as a primary phenomenon to study in meditation research, as well as in clinical work that utilizes contemplative techniques.

Shinzen is helping to create complex algorithms that will program a bot to teach meditation in a condensed and targeted way to accompany the ultrasound, thus streamlining the whole process with the aim of scaling it to reach large numbers of people. In Shinzen's view, the bot can become not just a mindfulness coach but also a mindful communication coach. As people communicate more mindfully with the bot, the bot will become more aligned with its user, providing free lifelong expert guidance in mindfulness. Pairing that with the ultrasound would offer access to deep practice for anyone ready to give informed consent. The result would be mainstream access to sophisticated bespoke practice at extremely low cost, which could have a historic effect on public health.

What do I think, you might ask.

As a teacher of meditation teachers, I'm attached to the idea of teaching meditation as a relational art that can't be replicated by an algorithm or a bot. As a meditator, I have both a reflexive suspicion of grandiosity and a commitment to open-mindedness and curiosity. As a fellow citizen of a world on fire, I'm open to any methods that reduce suffering and increase wisdom and compassion.

Putting Equanimity Front and Center

I first met Dave Vago years ago at a mindfulness conference in Worcester, Massachusetts, where he went far out of his way to give me a ride into Boston. We run into each other a lot at conferences, and he is always the same: warm, open, smiling, friendly, interested in the person he's talking to, and surprisingly unassuming for a neuroscientist with a lengthy list of prestigious credentials. His bio says it all. He's "on a mission to alleviate suffering and improve well-being through investigating connections between the mind, brain, and body." For my part, I find it interesting that here is yet another youngish parent (yes, even after logging in years as an academic researcher) who has worked with and been influenced by Shinzen Young, with whom Dave collaborated at Harvard in 2012 to explore the neurobiological mechanisms of mindfulness.

Most important, for our current interests, is the fact that he is an advocate of the importance of placing equanimity in a prominent place in research and was one of the authors of a consequential paper frequently cited as a cornerstone in researching equanimity as an important outcome of meditation practice and other contemplative practices. To begin with, Dave noted that positing equanimity as a key outcome of contemplative practice came out of years of a group of researchers working together as the Mindfulness Research Collaborative, which he said "met regularly, meditated together, shared data, and worked collaboratively on a number of theoretical papers." Eventually, equanimity ended up being the next big concept that this group of thirteen researchers wanted to write about "from cognitive, psychological, and neurobiological perspectives." They felt it was important that such a key concept needed to be clarified in scientific terms and more readily become a subject of research.

A principal focus for these researchers was on the *recovery time following an emotional stimulus*, known in the literature as "affective chronometry." Building on a framework pioneered by Richard Davidson in 1998, it refers to the temporal dynamics of emotional responses, specifically: how long they arise, persist, and dissipate over time. Vago and his colleagues posit that a primary signature of equanimity is "more rapid disengagement from initial emotional response and faster return to baseline."

They offer the example of responding to a loud noise. While there may be the expected initial startle response, with equanimity, "it is less likely to be followed by anxiety and the person will quickly return to their regular state." Equanimity, they write, "does not necessarily entail a complete lack of physiological response to emotional challenges." Such a lack of response, in fact, is usually considered "a mark of psychopathology." Vago and his fellow authors clearly state that "mere indifference is very distinct from equanimity." And that completely accords with my evolving understanding and what I am learning from the spiritual teachers I've spoken with about how *they* understand equanimity.

Groundbreaking research (reported in 2013) had already been conducted in Richie Davidson's lab at the Center for Healthy Minds, led by Antoine Lutz, that focused on responses to *physical pain*. It concluded that advanced meditators not only didn't numb their response to painful stimuli, they also recovered to baseline more quickly. Dave Vago and colleagues, among others in the field, are inspired to extend that to emotional pain.

The diagram below, adapted from figure 1 in the paper, illustrates the difference between the equanimous response (the solid line indicating rapid recovery); the perseverative (or prolonged) response; and the "blunted" response of someone showing low overall or apathetic response to stimuli.

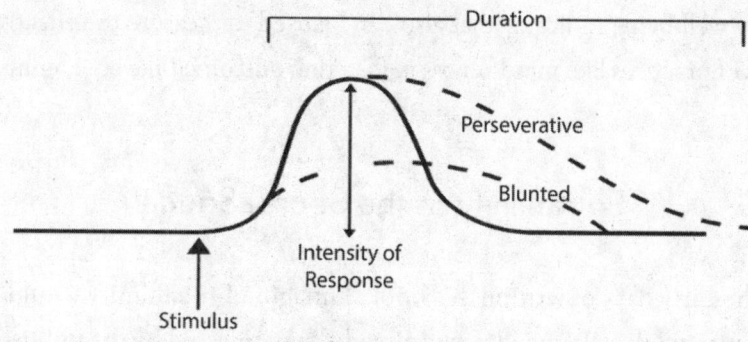

As Dave said in our conversation, the point is not to *feel less*; it is to actually *feel more* and yet recover more quickly—thereby increasing tolerance to challenging stimuli. In the paper the authors give an example of the anger that might emerge when a driver suddenly cuts in front of us in a dangerous way:

> Mindful of the arising of anger, we may (or may not) be able to maintain a state of equanimity and watch the experience of anger unfold and pass on its own accord. In the absence of equanimity, the experience of anger may become overwhelming or persist for a significant amount of time after the situation is over.

Dave's colleague Sara Lazar, another author on the paper, has taken the scenario one step further: After the car swerves in front, and your brain has responded as expected, and once you've reached safety, can you sit down and meditate again? Unfortunately (or rather fortunately), that is one study that would not get past the committees that ensure the safety of human research subjects!

In a related paper, Dave emphasizes that our well-known bias toward giving more attention to threat-related stimuli than neutral stimuli, while nonetheless adaptive, also can result in unhealthy attentional habits. Therefore, measuring the recovery time of our emotional responses can provide a key well-being indicator that also can be used to measure the effectiveness of practices like mindfulness in fostering outcomes like equanimity.

Equanimity as the Secret Sauce

In the early days of writing this book, I imagined equanimity would be a pretty tough sell. The idea had already been rejected by the publisher who invited me to write a follow-up to my first book, and nobody seemed to know or care what I was talking about outside the relatively small world of dharma devotees. Thanks to Google Scholar alerts, I've been getting pinged whenever an article is published that includes a reference to equanimity, and I read them all, so I've had my ear to the ground.

Imagine my surprise when I heard David Creswell, research psychologist at Carnegie Mellon, report in a podcast interview that his lab had pivoted from studying mindfulness to studying the "buzzy" field of equanimity. Wait . . . What?

Buzzing with excitement myself, I reached out to David and we had a great time geeking out about equanimity together. In spite of the fact that he was also planning to write a book about the subject, David was generous with his time and information. He said:

> I am completely zealous about equanimity. The more I look at the world as it is, the more convinced I am that equanimity stands alone as a quality worth developing, whether or not we bring it with mindfulness. I also believe that equanimity can be cultivated apart from mindfulness.

David smiles a lot and looks you straight in the eye when he talks to you. He's full of boyish enthusiasm about the potential of equanimity to both improve personal and societal well-being as well as its potential to be scaled. His lab has done research comparing mindfulness with and without explicit equanimity training. The main difference was in how the instructions were offered. In the equanimity conditions, the instructor used language laden with specific attitudinal "flavors" of acceptance, openness, curiosity, non-reactivity, and non-judgment.

According to David, when you remove this language, you "wipe out a lot of the benefits of mindfulness interventions." These include the stress-buffering effects on biological stress reactivity, reduced loneliness, as well as improved social interactions, "positive" emotions, and general well-being.

"We're no longer asking *whether* mindfulness interventions work," David tells me. "We're asking *which* mindfulness interventions work better than others. And when you put equanimity skills front and center, the answer seems to be a provisional 'yes, this works better.'" He went on to sum up:

> It feels like we're starting to figure out the secret sauce behind a lot of mindfulness programs. Millions of people try meditation and feel various kinds of discomfort and think they're doing it wrong. Not so. In fact, moments of craving, aversion, distraction, sleepiness, agitation: these offer the resistance training we need to build equanimity skills.

David becomes especially enthusiastic when he talks about craving as a doorway to equanimity. In a 2013 study conducted by his lab, instead of suppressing cravings, smokers were encouraged to engage with their cravings in an open and curious manner, focusing intently on their craving responses rather than resisting them. Compared to other prompts, such as *respond naturally, do whatever you want,* or *actively resist,* this method of accepting and allowing—or what David refers to as an "equanimity prompt"—significantly reduced their self-reported cravings.

Capping our conversation by invoking a classic Shinzen aphorism, David said, "With equanimity you can have your cake and eat it too . . .

The pleasant moments are more fully savored and the unpleasant moments increase our ability to be non-reactive and to learn how to manage challenge."

Measuring Equanimity

As the great mathematician and physicist Lord Kelvin said, "When you can measure what you are speaking about, and express it in numbers, you know something about it." The corollary to that has been paraphrased as "If you can't measure it, it's not science." While we may know equanimity when we feel it and see it in others, researchers need measurements. It is worth noting, though, whenever we talk about measuring something, we have to appreciate what I said about labels at the beginning of the chapter. When we're talking about aspects of the mind and behavior, the labels are a bit on the fuzzy side. Also, as philosophers like Evan Thompson and Michel Bitbol have pointed out, scientific research abstracts from everyday experiences and these abstractions can create blind spots and distortions. While measuring equanimity will never be like measuring a cup of flour, it still can bring value.

We've already seen one model for measuring an important aspect of equanimity: the measurement of the recovery time in response to emotional stimuli. No researcher is ready, though, to claim it as a comprehensive scale for equanimity. Dave Vago does think it's important, however, for any measurement of equanimity to distinguish it from apathy. He would be the first to admit that equanimity measurement is in its infancy, at best, and his coauthors noted that the scales they reviewed "lack a theoretical framework and reveal the lack of a common agreement for the construct of equanimity."

More recently, a sixteen-item assessment tool, ES-16, has been developed that was published in the journal *Mindfulness* in 2021 and is available online. The sixteen questions divide into two subscales: "experiential acceptance," meaning how much do you resist your thoughts, feelings, and bodily sensations; and "non-reactivity," how readily do you react to events or "inhibit a previously learned response"? Also, Joey Weber and Michelle Lowe, both of the University of Bolton, Greater Manchester, have developed the Equanimity Barriers Scale, a self-report measure that aims to assess the types and intensity of the obstacles that stand between us and greater equanimity.

The most rigorous research and development, though, comes from Catherine Juneau, who recently returned from France to her native home in Canada to take a position as a postdoctoral fellow at McGill University in Montreal. In our conversation, she shared some of the passion she brings to her research into equanimity. In fact, every researcher in this chapter seems to be driven by a strong personal motivation to develop their own equanimity and to see it spread more widely.

In Catherine's case, a virtual train wreck of circumstances intensified her interest in the subject: she was finishing her doctorate when COVID-19 hit; she became pregnant with her second child just as her marriage was heading toward divorce and her father passed away; and she moved halfway across the world to take on a postdoctoral position. She'd begun meditation practice prior to this period, but naturally she found it extremely difficult to find the time and energy as a single parent beginning an academic career. She needed to find equanimity wherever she could. Gradually, she began to make some personal discoveries that emerged in concert with her thesis research. As she wrote to me in an email:

The more my thesis progressed, the more I discovered mindfulness, the more I realized we could see things differently, and that it was possible to disengage from our thoughts and emotions. The initially very abstract concepts surrounding mindfulness were slowly making sense to me.... I understood that change was slowly taking place within me, but it was difficult to put it into words. So outside meditation, in my everyday situations, the term "non-reaction" seemed relevant, but it didn't feel like "*not* reacting." Rather, the *automatic* nature of my reaction diminished or even disappeared. It was as if I were in control, and yet without that control requiring any effort... not reacting automatically to everything my partner said saved me from a lot of difficulties. This wasn't always the case, of course, but over time I paid more attention to putting my energy into (re)building my life than defending my ego.

This kind of personal feel for equanimity must have been a huge aid in building a numerical scale to measure it. The scale that Catherine and her team developed is called EQUA-S. It's based on a set of questions that aim to measure two dimensions of equanimity: an "even-minded state of mind" and "hedonic independence," each of them drawn from existing research. An example of a question from the even-minded set is "I am not easily disturbed by something unexpected." From the hedonic independence set: "When I look forward to doing something pleasant, I can only think about that." (Responding in the negative would represent *more* "hedonic independence.")

Hedonic independence sounds complicated, but what it implies is fairly simple: being less tied to finding fulfillment in seeking pleasure and excitement, which Catherine illustrates well in discussing her own experience of realizing a kind of pleasure addiction:

> I've always had a very conflicted relationship with pleasure, and I used to envy people who managed to "control" themselves by putting their long-term goal ahead of excitement. As for me, I had the impression that I only existed on an emotional rollercoaster, that my long-term goals were only interesting if they were exciting and brought me constant gratification.

In a conversation with Richie Davidson, he emphasized that while there is clearly value in self-report measurements, they rely on the trustworthiness of participant's perceptions of their own experience, and much work remains to be done before there is a physiological measurement of equanimity. Catherine Juneau and Dave Vago agree. If and when that day comes, I may be able to walk into a lab like Jay's and have data showing how equanimous my responses have been over a period of time.

As I consider what I've seen and learned about the state of equanimity research, the approach to measuring equanimity that accords most closely with my understanding is *to measure how well one recovers from your initial reaction,* as we saw in the chart on page 79. If we're asking someone whether they remain calm no matter what happens, is that realistic, or even desirable? The more realistic, the more *human* question is whatever occurs, how quickly do you return to equilibrium? To any neuroscientists or psychologists reading this, let me know if you're inspired to work on developing *that* measurement. Inquiring minds would love to know.

Postscript

One final note on neuroscience research in the equanimity arena bears mentioning. The Buddhist tradition includes a series of progressive states of deep concentrated absorption and tranquility, known as the *jhanas*. The fourth of the eight *jhana* states is usually translated as *equanimity*. While this deep state, sometimes described as trancelike, is labeled as such, this form of "equanimity" seems a bit different from the everyday equanimity that emerges out in the world. Nevertheless, *jhana* practitioners believe that entering these states can ultimately have beneficial effects over the long term, and meditators are being studied in a lab at Harvard who claim to be achieving these states. Matthew Sacchet, a neuroscientist and founder/director of the Meditation Research Program, founded in 2022, within Harvard's Department of Psychiatry at Massachusetts General Hospital, is conducting the research using a very sophisticated fMRI machine to map the neural correlates of *jhana* meditators. If you're interested in this line of research, I encourage you to refer to the publications listed in the endnotes for this chapter.

Chapter 5

THE PSYCHOLOGY OF A BALANCED MIND

If you set up feeling good as what you have to be feeling, then you are just creating another thing you are failing at.

—Mark Epstein

For a long time the field of psychology, bound by the legacy of Freud, focused mainly on pathology. Then, in the mid-twentieth century, humanistic psychology came along, partly as a reaction to both Freudian psychoanalysis and behaviorism. The humanists revived ideas from the Stoics, like human flourishing, living a life of meaning, and individual agency.

When the field of positive psychology emerged from these ideas, it increased the backlash against Freud by focusing on the cultivation of so-called positive emotions. Martin Seligman, avowed father of positive psychology, developed a counterbalance to the pathology-based psychology canonized in the *Diagnostic and Statistical Manual of Disorders*, the (in)famous *DSM*. He and colleagues cataloged virtues that had been identified throughout world history to celebrate what ought to be cultivated, in their view.

I've always taken issue with the idea that emotions are either positive or negative. At the heart of both equanimity and mindfulness, for example, is the capacity to open to the full range of experience without reactivity, without putting emotions into categories of liking or disliking, good or bad. So I was not surprised when a backlash emerged against positive psychology beginning with author Jack (a.k.a. Judith) Halberstam, who coined "toxic positivity" in their book *The Queer Art of Failure,* back in 2011. This idea was picked up by a number of psychologists, particularly Susan David, whose TED Talk on the subject got more than ten million views. Her insights reveal the tyrannical trap of staying on the sunny side, as so many aphorisms beseech us to do:

> Toxic positivity is forced, false positivity. It may sound innocuous on the surface but when you share something difficult with someone and they insist that you turn it into a positive, what they are really saying is, "My comfort is more important than your reality."

In her bestselling 2022 book, *Bittersweet: How Sorrow and Longing Make Us Whole,* Susan Cain describes the "tyranny of positivity" as a cultural directive that says "whatever you do, don't tell the truth of what it's like to be alive." Likewise, the late social critic Barbara Ehrenreich, in *Smile or Die,* recounted her experience of breast cancer treatment and railed against the health care establishment—and her fellow patients—for treating her anger and sadness as unhealthy and dangerous. What was expected was a surface-level cheeriness that amounted to no more than suppressing emotions we've been taught are uncomfortable.

"Positive" and "Negative" Emotions

In many ways, attributing positivity or negativity to emotions (what psychologists call valence) directly contradicts the spirit of equanimity: Meet all experiences with openness and impartiality. It's exactly where a lot of the trouble begins. Once we decide something is *positive*, we tend to cling to it and want it to continue, in spite of knowing that everything changes. The more attached we are, and the more things change, the more we suffer and the less available we are to what is actually happening *in this moment*.

Conversely, having decided something is *negative*, we don't want it and tend to suppress it or project it out onto the world, creating countless problems for ourselves and others. Turn on the news and you'll see incontrovertible evidence of the tendency to polarize. (Not surprisingly, *Merriam-Webster*'s word of the year for 2024 was *polarization*.) The news media feeds on the emotions that emerge from increased polarity—and there's been lots to feed on in the past few years. To wit: Vanderbilt University calculates an American "Unity Index." With 100 representing unity, the index was a disturbing 46.48 at the end of 2023. Around the time of the 2020 election, it was at an all-time low of 40, compared to 72 in 1991, after the Berlin Wall had come down. A big contributor to the recent low scores is an increase in people who declare themselves as "strongly" at one or the other poles of the political spectrum.

When we treat emotions as polar, we end up with polarity, as we overemphasize other people's negativity. Purely and simply, emotions are not polar. They're not positive or negative. They just are; they're part of being human, and they all have a function. I learned this directly from Paul Ekman, when I worked on the "Cultivating Emotional Balance" study I mentioned in the prologue. Emotions contain critically important data. They signal what matters and provide much of the depth and richness of being human. I also learned from Paul—and had discovered for myself through meditation practice—that emotions are fleeting. It is their nature to arise and pass away quickly.

Harvard neuroscientist Jill Bolte Taylor was a postdoc doing brain mapping when she experienced a severe brain hemorrhage. As she lost and slowly regained brain function, she was uniquely positioned to study her own experience through the lens of neuroplasticity. One of her many insights was what she calls the ninety-second rule, which encapsulates the fleeting nature of intense emotions. According to this principle, the physiological response to an emotion lasts approximately ninety seconds. Beyond this time frame, the perpetuation of the emotional state is driven by our thoughts and interpretations, which in turn elicit more emotions and even emotions about our emotions.

Emotions tend to get "stuck" when we either suppress them or ruminate on them, which not only damages our physical and mental health but also undermines the quality of our relationships. Some of us grew up in families where emotions couldn't be expressed, and they got buried in a mysterious black box we didn't dare open.

Unfortunately, most of us were raised with a false emotional dichotomy: either act them out or suppress them. To make matters worse, emotion suppression was glorified as something dignified, heroic, mature, strong, and desirable. Tragically, physicians were typically taught to suppress and deny emotions in order to cope, leading to burnout and countless physical and mental problems, not to mention patient dissatisfaction.

In addition to door #1 (act out or ruminate) and door #2 (suppress, repress, avoid, or deny), there is a door #3: equanimity. *Feel it but don't feed it.*

It's completely possible to feel everything—feel all the feels—and still survive and function. In fact, functioning improves because of all that beautiful energy flow in the system. Jainish and Prittesh Patel expressed it well in a 2019 paper on repressing emotions: "It is important to acknowledge that feelings and emotions are not responsible for health disorders and sicknesses. Rather it is the protracted reliance on self-defense against the expression of emotions and feelings that creates the tension required for the disease to thrive."

Needless to say, there are many factors that create the conditions for diseases to thrive and some, such as genetic predispositions, are beyond our control. However, allowing emotions to flow unimpeded through the system is one that we *can* influence, and on closer inspection emotions aren't as scary as they might seem. The more we avoid them, the more threatening they appear. Turning toward emotions can feel like pulling back the curtain on the Wizard of Oz or discovering the tiger is made of paper. I still have to remind myself of this when my mind insists that I will fall apart if I allow myself to feel a challenging emotion. In a study published in March 2023 in the journal *Emotion*, researchers found that people who habitually judge negative feelings—such as sadness, fear, and anger—as bad or inappropriate have more anxiety and depressive symptoms and feel less satisfied with their lives than people who generally perceive their "negative" emotions in a positive or neutral light.

For almost thirty years I facilitated psychosocial support groups for cancer patients. Cancer can be cruel, and often the treatment is even more brutal than the disease itself. Our groups were free of charge, and people came from many socioeconomic backgrounds and exhibited a wide range of diagnoses and prognoses. It was our job to make room for everyone. One of the more heartbreaking, though not uncommon, strategies people would use to manage their "negative" emotions was not allowing themselves to get too happy or too sad. This emotional blunting was understandable but tragically misguided. Not only were "negative" emotions being suppressed and causing even more trouble to overly taxed nervous and immune systems, people were pointlessly denying themselves the opportunity to fully savor whatever moments of delight and happiness were available to them. A lose-lose.

SUPPRESSING OUR EMOTIONS IS *NOT* EQUANIMITY
In Fact, It's Bad for Our Health

A joint study from the University of Texas and the University of Minnesota, partially funded by the US Army—published in Social Psychological and Personality Science—*found that* **subjects who did not acknowledge their emotions became more aggressive**. *Suppressing the emotions made the emotions stronger.*

A study by the Harvard School of Public Health and the University of Rochester, published in the Journal of Psychosomatic Research, *showed that* **people who bottled up their emotions increased their chance of premature death** *from all causes by more than 30 percent with their risk of being diagnosed with cancer increasing by 70 percent.*

A 2019 literature review, published in the International Journal of Psychotherapy Practice and Research, *concluded that "it is clear that expressing one's true emotions . . . is crucial to physical health, mental health, and general well-being" while* **concealing emotions is "a barrier to good health."**

Admittedly, we walk a very fine line here, but an extremely important one when it comes to cultivating equanimity: Rather than limiting happiness or sadness to avoid the highs and lows, we can instead expand our ability to be with the greatest possible range of joy and sorrow—without collapsing, losing perspective, or taking our emotions too personally.

Participants in my cancer groups also struggled with how to talk about their emotional experience. They were afraid to either share how hard their week had been for fear of complaining, or how lovely it had been for fear of upsetting those who were struggling. Here again, we explored door #3: simply reporting. This was a novel idea to many people: "You mean I can actually tell you how much pain I was in without complaining?"

Labeling emotions as positive and negative is just one way language plays a big role in shaping how we relate to experience. Even the subtle wording shift from *positive* to *pleasant* and *negative* to *unpleasant* can begin to loosen the reflexive tendencies to cling and push away. In that light, it's interesting to begin to pay attention to the adjectives and adverbs we use. Do they tend to exaggerate for effect? To minimize?

I recently listened to an interview with the wonderful meditation teacher Adyashanti in which he referred to the *tragic* dimension of life. What if we substituted the word *poignant* for *tragic*? For me, *tragic* triggers a feeling of collapse, whereas *poignant* is more about feeling moved. Would you call the weather tragic? What does that word add to the experience of weather? There are ways of expressing emotions to both ourselves and others that tend to sabotage equanimity and are worth examining. For example: being flabbergasted, outraged, horrified, infuriated, and so on.

Trevor Noah did a comedy sketch about the word *flabbergasted* that made me laugh out loud and also made me stop and think. He talked about some key differences between white people and Black people and how white people are flabbergasted (and Black people aren't). It hadn't occurred to me before that being flabbergasted is a luxury of privilege. I started thinking about my own tendencies to hyperbolize and dramatize as I recount a story to myself or others. How much of my own suffering is manufactured either for effect or as a defense against boredom? It was revelatory to begin to look at my own use of hyperbole as a hidden signal of my good fortune.

TRY THIS

Report on emotions to yourself or others as you would on the weather. Do your best to neither exaggerate nor minimize. But if you do, don't be hard on yourself. It's good feedback and can help you become increasingly aware of your tendencies to manage your emotional life. Notice how language plays a key role here.

Psychology and Emotion Today

Although Western psychology has embraced mindfulness-based interventions for mental health, it has just recently (the past ten or so years) begun to explore the concept of equanimity. My investigations led me to four researchers who are independently exploring how to incorporate equanimity into contemporary psychology. Gratefully, they all agreed to speak with me, and, interestingly, for the most part they weren't aware of one another's work.

When I came across the paper "Judging Emotions as Good or Bad: Individual Differences and Associations with Psychological Health" from Dr. Iris Mauss, director of the Emotion & Emotion Regulation

Lab at UC Berkeley, I immediately reached out to the authors. I was grateful to hear back from Iris, who was happy to share her take on the latest research on emotion regulation and how the field is evolving, particularly regarding positive psychology. She shared how her recent research not only confirms the blowback from seeking only positive emotions but also may be part of a new trend in psychology that's embracing the importance of equanimity. At many points in our conversation, I found myself happily answering questions about my book and equanimity and forgetting that I was supposed to be interviewing her. Iris is one of those people whose natural curiosity and genuine interest in other people make her talk less and listen more. In spite of juggling being at home with young children and a demanding role in academia, she was as eager to learn from me as I was to learn from her.

"Until pretty recently," she told me, "emotion regulation was researched almost exclusively as something directed at decreasing negative emotions. As a culture, we have a tendency to think of negative emotion as something that needs to be avoided almost at all costs." In fact, her research shows that people who try hard to rid their lives of negativity or are overly focused on striving after the positive are not necessarily the healthiest people.

According to Iris, those people may well have more risk for depression and anxiety and less well-being in their lives. She goes on to say that equanimity seems to be in the zeitgeist now, as more people try to find composure while not dampening the zest for life. For example, Maya Tamir and her colleagues at the Emotion & Self-Regulation Laboratory at Hebrew University of Jerusalem have done extensive research that looks beyond the simple binary of positive and negative emotions, demonstrating that the context and functionality of emotions are as important as their positive versus negative valence. What Iris, Maya, and others suspect is most valuable for well-being is to experience the range of emotions, which segregating and redlining negative emotions doesn't allow for.

"People don't generally experience any one emotion purely without any other emotions," Maya says. "In fact, we and others have found that if you experience more multifaceted emotions—positive and negative ones together, multiple flavors of negative emotions together, even multiple flavors of sadness together—you tend to have better psychological health and lower levels of depressive symptoms. There's something almost magical about the simple idea that if you just change your relationship with your emotions, it deeply transforms the emotion."

The Second Wave of Positive Psychology

I was first alerted to the work of Tim Lomas, research scientist at the Human Flourishing Program at Harvard, by his article "Cultivating Equanimity Through Mindfulness Meditation." This study equated the idea of *decentering*—not taking our thoughts so personally—with equanimity and found that increases in decentering resulted in decreased anxiety and depression, particularly for men.

Tim Zoomed me in from his backyard somewhere near Seattle, while his young daughter napped inside. Tim spoke with a soft voice and a heavy British accent. I was amused to see a genuine pub dartboard on the brick wall of his house, framed neatly over his left shoulder. In reviewing the recording of our conversation, I was struck by how often he paused, tilted his head back and over his left shoulder, and closed his eyes as he let the question sink in and looked for a deeper answer. Most people are already answering questions before the question is fully asked, which is a pretty good indication that they know what they want to say.

Tim has also written extensively on the second wave of positive psychology. Originally spearheaded by Canadian psychologist and academic Paul Wong as Positive Psychology 2.0, this theoretical model is new enough that it remains relatively unknown, in spite of numerous articles in peer-reviewed journals. It wisely chooses to build on the work of positive psychology by engaging in a dialectical conversation rather than rejecting it altogether. Much like the work of Iris Mauss, Susan David and others, this second wave is questioning the false dichotomy of positive and negative emotions, preferring to use a more nuanced lens for examining emotions that

- cautions against the labels of positive and negative;
- tolerates paradox;
- sees emotions as complex phenomena that include darkness and light; and
- offers an evolutionary view that builds on, expands, and refines what has come before.

Take sadness, for example. Often characterized as a negative emotion (as we clearly heard from Barbara Ehrenreich), sadness can arise as *the first step toward a compassionate response to suffering*. Sadness can carry important information, such as how deeply we care about someone or something. It can also be a source of inspiration or significance. It can even harbor beauty. We cry the same tears whether we are moved by sublime beauty or sorrow. Sadness can open the heart's door to our shared common humanity.

At the same time, the pressure to be positive can cause harm by discouraging people from facing their situation. This results in denial of both the reality of difficult circumstances and the emotions that accompany them. Forced positivity can become a maladaptive coping mechanism.

Global Well-Being

What intrigued me even more, though, is Tim's work with the Global Wellbeing Initiative (GWI), which is part of the famous Gallup Poll and aims to reshape how we think about and measure well-being. In our conversation, Tim told me that his current work involves integrating the values of balance and harmony (both closely related to equanimity) into how we understand well-being. The GWI is partnering with a Japanese Foundation called Well-Being for Planet Earth to further develop less Western-centric standards of human flourishing. Not surprisingly, Western ideals around well-being focus on high-arousal positive emotions and life satisfaction; whereas, balance, harmony, and peace tend to be low-arousal states that are more highly valued in Eastern cultures. Tim explained further that he believes harmony and equanimity are not entirely limited to low arousal, which aligns with the way we're thinking about equanimity here. For example, he described flow states as potentially both calm and high arousal: "I can imagine someone in a state of quite high energy nevertheless feeling calm and a sense of equanimity, like in the midst of a sporting event or musical performance."

His work is having global impact and underscores Iris Mauss's instinct that the zeitgeist is shifting toward an appreciation of the quiet strength of equanimity. In the 2022 World Happiness Report, there's a chapter called "Insights from the First Global Survey on Balance and Harmony." His group has also consulted with the Organisation for Economic Co-operation and Development on their guidelines for assessing well-being, and they subsequently added a section on balance, harmony, and inner peace. Tim summed things up toward the end of our conversation:

> If we push just slightly—enough to open the door and point out the importance of these concepts—people take to them

quite readily. Though they've been overlooked somewhat, the picture is changing and people are quite receptive to the importance of balance, harmony, peace, calm, equanimity.

The Quiet Ego

Jack Bauer—author of *The Transformative Self*—is a psychology professor at the University of Dayton in western Ohio who has focused much of his research on the development of self-identity and personality, with an emphasis on life stories, meaning-making, growth motivation, and human flourishing. I learned from our conversation that he's also a bit of a "closet" Buddhist practitioner, having first encountered meditation as a teenager trying to relax his grip on the baseball bat to improve his performance. His interest in Buddhist philosophy proved a lot deeper than stress management and has quietly informed his professional and academic pursuits.

Along with his colleague Heidi Wayment, he's developed a psychological model of human flourishing called "the Quiet Ego" that includes many of the hallmarks of equanimity. It consists of four interconnected facets: detached awareness, inclusive identity, perspective-taking, and growth-mindedness. These four characteristics all contribute to having a general stance of balance and growth toward self and other.

Heidi and Jack began the conversation that eventually led to the idea of the Quiet Ego in the early 2000s, around the time of the Iraq War. As Jack told me:

> The political polarization was bad at that time. There were a lot of newspaper stories about how people can't even talk to each other. We immediately thought, "This is a noisy ego problem."

So that's how we framed it at the societal level. Now, since the Trump era, violent rhetoric and violent acts have become part of everyday political discourse. Once again, the noisy ego. It has ripple effects. White nationalism used to not have a voice; now it does.

Much like a fantasy dinner party with the most scintillating guest list, Jack and Heidi decided to host a conference on the Quiet Ego and invited researchers and academics who were doing interesting work in related fields. If you dream it, they will come. They came, the money came, and an edited volume came: *Transcending Self-Interest: Psychological Explorations of the Quiet Ego*.

Perhaps a silver lining to the degeneration of global political dialogue is the hunger for something better. Top researchers accepted the invitation to the Quiet Ego conference because they shared this deep longing for civil discourse and balanced, wise leadership. The clock is ticking on climate change at an alarming pace. At the same time, tribalism and identity politics are on the rise. As tribal conflicts continue to escalate, we recklessly squander precious resources we need to collectively address the future of our planet.

And what of our leaders? Politics today more than ever is full of noisy egos—including some that are deafeningly loud. Many, if not most, of us inevitably recoil from an ego that noisy, longing to advocate for another way of being in the world. It's not surprising, then, that this theoretical model is emerging *now*.

The field of psychology overall is beginning to ask different questions about what psychological health might really look like. Should we include the ability to balance our own needs with others' needs? Is humility a part of psychological health? Wisdom? Non-reactivity? All the things that Jack describes are directly related to equanimity. Tim openly acknowledged that both the Gallup Poll and Western psychology have been subtly colored by Western values around independence versus interdependence, arousal versus tranquility, humility versus pride in achievement, and so on. Now, it seems a shift is occurring in the direction of what most ancient wisdom traditions in the East as well as many Indigenous traditions around the world have extolled.

We're seeing—to the extent we're willing to face it—the devastating impact of denying our interdependence with each other and the planet. The Western ideals I was raised with—independence, individuality, winning at all costs, goal-seeking, and achievement—have allowed us to do many things, but at a great price. And the same could be said for the lens I was asked to see the world through in my clinical training to become a psychotherapist: abundant sex life, professional success, and "positive" mood. Interdependence, balance, solitude . . . equanimity—these were not front and center.

Since Freud bequeathed us the concept of ego, Western psychology has been interested in how to make the ego healthy and strong. This is in stark contrast with most Eastern philosophies, which are more concerned with the collective and embedding the "self" in an interdependent universe. Rather than impose an either/or dichotomy on the question of ego, Jack and his colleagues came up with an apt metaphor:

Acoustic volume goes from low to high or from quiet to noisy and can be measured in degrees along a spectrum.

A noisy ego is turned up *loud*. It's self-focused and constantly clamoring for attention. Like a child, it demands that its needs be met. A quiet ego is more mature, able to self-soothe and balance the needs of others with its own needs. It doesn't require constant praise and attention. Simply put, it is equanimous.

We can see the principle of a quiet ego reflected in many different wisdom traditions around the world. For example, in the Seven Grandfather Teaching seen in many Native American contexts, humility is one of the key virtues tied into the other six, as is well expressed by the Nottawaseppi Huron Band of the Potawatomi:

Humility is to know that we are a part of creation.

We must always consider ourselves equal to one another. We should never think of ourselves as being better or worse than anyone else. Humility comes in many forms. This includes compassion, calmness, meekness, gentleness, and patience. We must reflect on how we want to present ourselves to those around us. We must be aware of the balance and equality with all of life, including humans, plants, and animals.

The Abrahamic traditions (Judaism, Christianity, Islam) in both their mystical and more conventional forms, all contain notions of self-abnegation, a lessening of the self in order to allow a merging with the divine. Habīb Boerger, the Sufi teacher and scholar we met in chapter 2, shared a wonderful way of relating to the ego from a Sufi perspective: "One of the ways I think about the spiritual journey is a process of befriending the ego and the ego learning to trust. Then changing your relationship to the ego from being your boss to being your personal assistant."

And then... Some People Just Have It

One of the key distinctions in psychology is between a state and a trait: A state is momentary and conditional, and a trait is an ongoing feature. It's quite common to study, measure, and compare "state mindfulness" (perhaps induced by meditation practice) and "trait mindfulness" (how mindful someone is on a day-to-day basis). Based on my experience of the many people I have met in my life and work, I've come to feel that some people simply are endowed with a strong dose of the trait of equanimity, pretty much apart from any practice they might do to cultivate it.

Take my good friend Jon. Though we don't live in the same city, his significant other is one of my best friends, so we've spent a lot of wonderful times together. Although most people would describe Jon as laid-back and easygoing, he's one of the most engaged and caring people I've met. He's deeply concerned about politics, people, and the environment and puts his money where his mouth is. He is a vegan, a caring friend, a loving son, and, after retiring, he became actively involved in a nonprofit that creatively addresses the problem of deforestation in Latin America. Though he is low-drama, in no way is he indifferent to the suffering of the world.

A few years ago, while paddling out to surf in Costa Rica, Jon was attacked by a crocodile. It's a miracle he survived. The crocodile's first assault mangled Jon's ankle. He fought him off, but the crocodile came back for more, trapping Jon's entire head in his mouth, breaking his nose, and leaving gashes in his neck where the crocodile's fearsome teeth dug in. Astoundingly, he fought him off yet again. In the end, he lost his lower leg.

When I asked him about the experience, Jon reported the same extraordinary calm and determination that many people experience in the middle of a crisis, which no doubt helped him to survive. In Jon's case, though, I was taken with the fact that he was the same chill guy before, during, and after the crocodile attack. Once he made it home, he simply proceeded to get on with life, which included several surgeries and learning how to live with a prosthetic leg. As soon as he could, he was back doing all the sports he loves. (Well, almost. You can't really ski with a prosthetic ankle that doesn't bend.) He even did a yoga teacher training in Brazil.

The psychotherapist in me kept plying Jon with probing questions about PTSD. He didn't seem to have any. He had no phantom limb pain or any of the other skin problems that often plague amputees. I couldn't help wondering if these were all related to his temperament. For someone with natural equanimity, like Jon, very little time and energy is spent wondering "Why me?" and recounting with ever-greater drama the worst details of his ordeal to anyone who'll listen. All that energy seems freed up to allow his body to heal. He isn't caught up in endless rounds of replaying the scene and calling up the stress response. Jon's license plate reads "FG2BA" which stands for "fuckin' great to be alive."

I became more and more curious about equanimity as a personality trait in writing this book. I even asked Jon to talk to his elderly mom and tell me what he was like as a baby. Here's what Betsy said: "As a baby, Jon seemed quiet, cool, and calm. I don't remember him crying a great deal. He was easily calmed by me. He just seemed happy with life. He sees both sides of the picture. That's an ability not everybody has, and it's made his life very good."

Equanimity as a trait may also be found within one's cultural heritage. Kiliii Yüyan, whom we'll meet more in Part II, was raised near Seattle by his parents and his Nanai grandmother. The Nanai are an Indigenous nation from Siberia/Manchuria, and Kiliii exhibits traits that may be widespread in Nanai culture, as he told me: "My brother and I are extremely steady and calm under fire, extremely nondramatic, which isn't to say that we don't have fun. We're really interested in what will be effective, what is useful, rather than what will make *us* feel good. And we inherited a lot of pragmatism from my parents, who survived the communist takeover and fleeing to the US. Equanimity is a really pragmatic tool—the ability to be solid within yourself when all the world around you is raging."

Probably the biggest reason I'm writing a book about equanimity is that I feel I have too little of it (ask my daughter). At the same time, my husband of thirty-five years will testify to the fact that I'm more equanimous now than when he met me. Sure, age no doubt plays a part in that, but I've also been blessed with great teachers and exemplars who have introduced me to the many practices for developing equanimity that form the second part of this book.

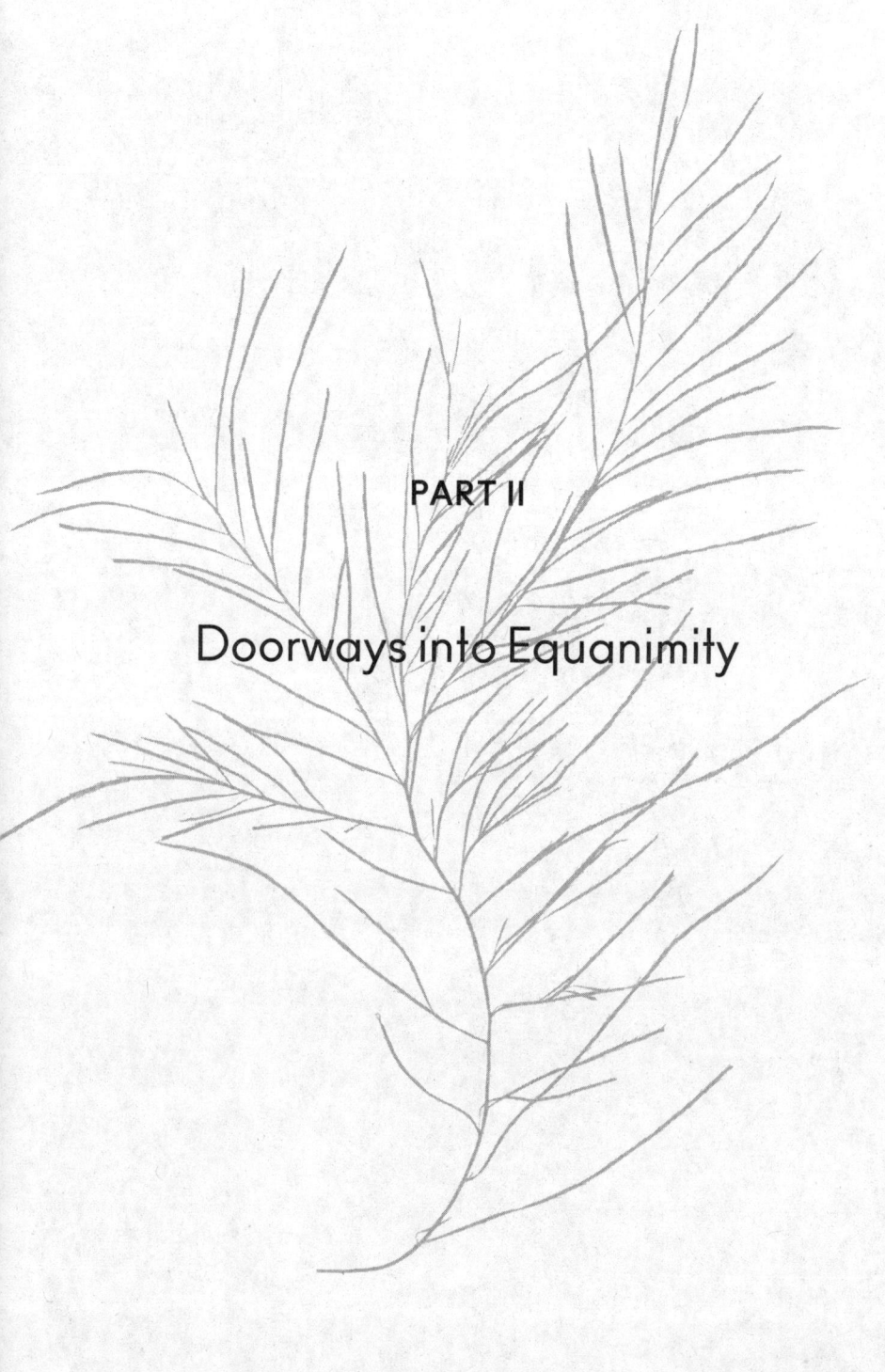

PART II

Doorways into Equanimity

Introduction to Part II

DOORWAYS INTO EQUANIMITY

In equanimity, every cell in your body
will generate sweetness.

—Sadhguru

In Part II we will explore ways to cultivate, access, and uncover equanimity.

As you learned in Part I, if you're already practicing mindfulness, you've also been cultivating equanimity. If you're *not* interested in, or have *not* been practicing mindfulness, the following chapters will offer exercises that stand alone for cultivating equanimity.

If you've never practiced meditation, these exercises will provide a wonderful foundation you can bring into virtually any meditation practice you choose—whether it's mindfulness or any other type of contemplative practice.

It turns out that moments of equanimity are closer at hand than you might imagine, and there are lots of different ways to cultivate and uncover it. Like mindfulness and other virtuous mental states, equanimity can't be forced. In some ways, these qualities behave like shy animals, only appearing when the conditions are right. Our job is to create the conditions for equanimity to arise. Just like any good skateboarder, surfer, or gymnast knows, balance and equanimity do not emerge from striving, tensing, or trying to control the outer world.

Equanimity is a relationship, a beautiful dance, between inner and outer, expansion and contraction, intention and surrender.

In this section, you'll see a wide range of techniques that represent different ways to invite greater equanimity. I want to offer you the fullest range of possibilities and also respect the reality that we're all different and that circumstances constantly change. There are many portals into equanimity, and there's no hierarchy of one method over another. I also encourage you not to take it personally if any set of instructions doesn't fit for you.

This is not a test. There are no grades, not even pass/fail. You're invited to experiment in the privacy of your own heart and mind, and to bring a light touch to these practices. And don't forget your sense of humor. It's one of your greatest allies in working with the mind and the deeply conditioned tendencies that are being gently uprooted through these practices.

For many years, I taught MBSR at a beautiful psychiatric facility that was part of Kaiser Permanente. We always had our silent daylong practice sessions on Sundays, when no one else was in the building and we could do walking meditation in the lush inner courtyard. Students were taught to slow down sufficiently to become mindful of every step they took, and many people chose to walk so slowly they could barely maintain their balance. I would often remind them that it was OK to laugh (silently of course) as they imagined an unwitting stranger seeing zombies stalking the grounds of the hospital. Laughter, as we'll see in chapter 13, breaks the spell of taking ourselves too seriously and becoming too tight for mindfulness to arise.

Likewise, if you notice yourself working too hard at being equanimous, humor is your friend. Maybe you've decided that equanimity has a certain posture or facial expression and you're unconsciously mimicking the way an equanimous person *should* look. This can become a welcome red flag that signals a near enemy of equanimity may be asserting itself. Ask yourself, "Am I fully able to feel what is happening in this moment or am I assuming a posture of looking 'chill' as a way to manage my feelings?"

For the skeptics out there, you might notice that some of these practices seem to contradict one another. We'll talk more about paradox in Part III, but for now, one way to assuage the desires and dictates of your logical mind is the elegant concept from Buddhism of "skillful means," *upāya* in Sanskrit. *Upāya* essentially "endorses" the teacher to adapt or tailor the teachings according to their audience. It was this very flexibility that allowed the Buddha himself to reach theistic audiences and for Buddhism to spread throughout the world. The teachings are understood as a means to an end, a finger pointing at the moon, and not anything to be attached to. When we approach these practices through a more utilitarian lens, it is easier to tolerate seeming contradictions.

Jack Kornfield shared a beautiful example of this from his days as a monk in the Thai forest monastery of Ajahn Chah. As a Western, Ivy League–educated young man, questioning and doubting had been inculcated into him. One day he heard his beloved teacher give contradictory instructions to two different students and challenged him with a "gotcha" moment. Ajahn Chah simply replied:

> My way of teaching is very simple. It is as though I see people walking down a road I know well. To them the way may be unclear. I look up and see someone about to fall into a ditch on the right-hand side of the road, so I call out to him, "Go left, go left."
> Similarly, if I see another person about to fall into a ditch on

the left, I call out, "Go right, go right!" That is the extent of my teaching.

In this way, both the stillness of the mountain and the constant course correction of a gyroscope can be opposing but equally valid evocations of equanimity.

Another way to think about the practices in this section is to use the analogy of a camera, which can take equally beautiful pictures with a wide range of different lenses. Likewise, our "lens" for equanimity can be useful whether we zoom in on moment-to-moment experiences or take the widest possible view in order to gain perspective.

Many of the following chapters will feature meditations. Feel free to follow along with the exercises, access audio versions via the QR code on page 7, or simply read through them first and come back whenever the time feels right. If you're following along with the written instructions, please remember to take your time with each of the phrases you're contemplating. There's no rush.

Whatever you choose to do, research shows that writing down your reflections helps you remember, consolidate, and better extrapolate what you learned to other areas of your life. I encourage you to experiment with writing down any insights that arise as you try on these different practices.

Whenever you engage with practices such as these it's wise to resist the temptation to "should" yourself. This habit runs deep for most of us and is a perfect set-up to create resistance. It can be helpful to remember that there's no parent, teacher, or authority hiding anywhere in these pages who will be measuring your participation against any kind of yardstick. (When you're practicing, there is also no big brother or mother or evil internet conglomerate tracking your eye movements—let's hope.) You are both free and encouraged to dip your toes in and out of the refreshing waters of equanimity and proceed if the temperature feels right for you.

Although you can't demand equanimity, there are several things you can do to create the conditions for equanimity to arise. First and foremost, you can bring an attitude of warmth, openness, acceptance, and curiosity toward your own experience. You can also learn the territory, or "inner" geography, of equanimity—what it feels like—so that you can recognize moments of equanimity when they arise, savor them, and learn how to replicate them. You can incline the mind toward equanimity by reflecting on phrases that evoke wisdom and clear seeing. You can read stories that inspire and illuminate the quality of equanimity and you can engage in specific meditation practices that evoke equanimity. All of these are available in the chapters just ahead.

Outside of this book, you can engage with a spiritual path or philosophical tradition that will help you to experience and trust a larger sense of the world beyond your ego identity. And you can follow the Buddha's advice, who urged us to seek out and spend time with wise and equanimous friends. In an article in *Tricycle* magazine, the great Burmese meditation teacher Sayadaw U Pandita suggested we could cultivate equanimity by avoiding friends who are overly dramatic and by "choosing friends who stay cool."

Perhaps even more important, you can't inflict equanimity on others. I know, I know, it's very tempting. My students frequently ask me how they can get other people to be more equanimous, particularly their children and partners. The answer is, you can't. Please don't try. Cultivate your own equanimity; it may rub off, but don't base your own equanimity on waiting for such an outcome. However, if you really want to make sure a person absolutely cannot achieve equanimity, next time they fly off the handle, you could say something like "Aren't you reading a book about equanimity?"

Chapter 6

SHIFTING PERSPECTIVE

Since my house burned down
I now have a better view
of the rising moon.

—Mizuta Masahide

This haiku captures, as so many do, a lovely equanimity that emerges from shifting perspective—taking a bigger view. In fact, in my experience, there is no prose or poetic form that better captures equanimity than haiku. The strict constraint of seventeen syllables leaves no room for drama or hyperbole, allowing for the most successful haikus to convey quiet, profound truths with warmth, humility, and lightheartedness. They can very quickly expand the mind.

To practice shifting perspective, in this chapter I'm asking you to try three thought experiments, which involve bringing to mind situations that are troubling you, then doing some reflections that support the perspective and cultivation of equanimity. Please don't be deceived by the simplicity of these little experiments. They are simple to do yet powerful enough to change the world. After all, $E=mc^2$ came from a thought experiment.

The following three exercises were inspired by a talk on equanimity from my beloved meditation teacher Joseph Goldstein. His insights motivated me to develop a workshop that puts these ideas into practice, providing a straightforward and effective route to cultivating equanimity. It's important to approach these experiments with the appropriate frame of mind, so it may help to look at the box of caveats that accompany this chapter on page 125.

Thought Experiment #1: Nothing Lasts Forever

> That which blossoms falls, the way of all flesh
> in this world of flowers.
>
> **—Author Unknown**

This first reflection involves applying the perspective of impermanence.

Begin by asking yourself if you can think of anything in your experience that is permanent, that never changes. For the sake of this exercise, let's leave out the metaphysical realm. In the material realm, is *anything* permanent?

Though we may know intellectually that everything is impermanent, we don't necessarily behave or make choices based on this understanding. Do all relationships end in separation? Is that really true? If you really believed this, would it change anything about how you relate to others? About appreciation? When you're in pain, do you remember the truth of impermanence? Are there situations when you forget this truth?

One of my favorite children's books is called *That's Good! That's Bad!* by Margery Cuyler. I often read parts of it out loud to my classes on equanimity. It's about a little boy who went to the zoo with his parents and they bought him a shiny red balloon.

"It lifted him high up into the sky. Wow! Oh, that's good. No, that's bad."

It turns out the balloon drifted for miles until it came to a hot steamy jungle and broke on the branch of a tall prickly tree. "Oh, that's bad. No, that's good."

The little boy fell into a muddy river. He climbed onto the back of a hippopotamus and he rode to shore. "Oh, that's good. No, that's bad."

Ten noisy baboons were squabbling in the grass by the river and they chased the little boy up a tree until he was out of breath. "Oh, that's bad. No, that's good."

You get the picture. Everything changes and things often aren't what we think they are. It is the nature of all conditioned things to change.

PRACTICE

Bearing this in mind, take a moment right now to recall a situation that's troubling you. It's wise to begin this type of exercise with a situation that feels manageable, no higher than a five on a scale of one to ten. It's rarely helpful to intentionally trigger overwhelm. For this exercise, choose something smaller than the global climate crisis, for example, and more personal, but current.

Bring it to mind with as much detail as possible.

Notice how it feels in the body and mind as you make it as vivid as you can.

See if you can find the feeling (or feelings) that are the most difficult: fear, anger, shame, grief.

Whatever they are, reflect on other times you've felt that way.

What happened to those feelings?
Does any feeling stay forever?
Reflect on this current situation in the light of impermanence and notice if anything shifts. Bring an attitude of openness and curiosity, without forcing any particular outcome.
If you find it helpful, journal about your experience.

Thought Experiment #2: Don't Take It So Personally

The second experiment will explore a different shift in perspective—focusing on the distinction between taking things personally and perceiving them as impersonal. Given my extensive experience with the former, I feel especially equipped to guide this exercise.

A renowned Chinese parable poignantly demonstrates the unnecessary suffering caused by mistakenly interpreting impersonal situations as personal. I have turned to this lesson countless times to remind myself that most situations that cause me pain are not as personal as they may feel in the moment.

A man was rowing his boat upstream on a misty morning. As he looked ahead, he saw another boat coming downstream, heading straight toward him. He clearly had the right-of-way but the oncoming boat would not yield. At first, he began shouting at the boat to steer away, growing increasingly angry as the boat continued on its direct course toward him. His shouts turned to rage as he screamed to avoid a collision.

When the fog lifted and the boat came close, he saw something very clearly: The boat was empty—it had simply drifted downstream with the current.

PRACTICE

Bring to mind another situation that is troubling you, perhaps one in which someone was rude to you, insulted you, disrespected you, or hurt your feelings. Again, choose something that feels manageable, not overwhelming.

Notice how it feels in the body and mind as you make it as vivid as possible.

See if there's a way to interpret the situation as impersonal—you're not being singled out, and there's no one to blame.

Is it possible there are causes and conditions that gave rise to this circumstance that are impersonal? That you may never fully understand?

Is there any level on which this might be true?

When you try on this perspective, what happens to your distress/anger?

Take a few minutes to write your reflections if you find that helpful.

There! You've now had a taste of "no-self," the Buddhist concept that challenges everyone so much. It's pretty easy to grasp in the context of not taking things so personally. No self, no problem. The idea isn't that the felt sense of a "self" will ever disappear, nor should it. (How could you ask for a cup of tea?) Rather, it's about refraining from imputing more power, permanence, and solidity to this sense of self than it actually merits.

This idea is also not limited to Buddhist teachings. We find it in many major religions. The Gemara is a key Jewish text, forming a core of the Talmud. This is one of many places the idea of no-self can be found in Jewish mysticism:

Rabbi Yochanan taught: The words of Torah cannot be fulfilled except by one who makes himself as if he does not exist.

Judaism also contains the idea of *Bitul Hayesh*. This Hebrew phrase, meaning "nullification of existence," is a Kabbalistic concept involving the sublimation or diminution of one's ego in the presence of the Divine. It's the idea of minimizing one's self-importance to become more receptive to spirituality and divine will, acknowledging that the self is not the center of existence.

In Sufism there is the similar concept of *fana*, which refers to the annihilation of the ego or self. *Fana* represents a spiritual state where the self dissolves, and the individual's consciousness is completely absorbed in divine presence. This state is characterized by the loss of individual identity and a profound realization of one's unity with God.

The great Sufi poet Rumi wrote in his *Masnavi*:

Don't become united with yourself at every moment,
> like a donkey stuck in the mud.
> You see a mirage from a distance and you rush;
> you fall in love with your own discovery.
> If the self tells you to fast and pray,
> It's but a trickster, hatching a plot against you.

Similarly, Christian mysticism often speaks of becoming one with Christ, where the individual self is subsumed into a greater divine reality. Mystics like St. John of the Cross and St. Teresa of Ávila discussed losing themselves in the love of God, which can be seen as a transformation or dissolution of the ego-centered self in the union with divine love.

The concept of no-self is generally taught within Buddhism without reference to a supreme divine presence, often resulting in the assumption that it is an atheistic religion. A more accurate characterization would be non-theistic, aligning it more closely with contemporary Western scientific perspectives. Stanford biologist and neuroscientist Robert Sapolsky extends this idea to its utmost implications in his latest book, *Determined*. Drawing extensively from the hard sciences, particularly biology, he argues that there is no "man behind the curtain," no conductor orchestrating the drama of our lives. In fact, whether you lose the sense of "I" through a flow state, a connection with the divine, or an experience of non-dual reality through meditation, the idea of a "self" choosing (exerting free will) becomes moot.

Thought Experiment #3: Big Picture

This final exercise introduces the perspectives of space and time.

In states of worry, pain, or self-absorption, our world may feel constricted, and our perception of time and space can feel oppressively narrow. It's important to note that these thought experiments aren't necessarily designed to teach you something new. Instead, they acknowledge that perspective is something we gain, lose, and regain repeatedly. This particular reflection invites you to consider how adopting a broader perspective has the potential to alleviate suffering in an instant. This shift might be as straightforward as a figure/ground reversal, or an orthogonal rotation in consciousness, as discussed in chapter 1. However, like all exercises in this chapter, it can't be forced. You're encouraged to experiment and discover the effects for yourself. A mentor once told me, "You're not a good guy if you do and you're not a bad guy if you don't." You may experience immediate relief, or you might find you're not receptive at the moment. You could try again later or apply the exercise to a less triggering situation. Whatever the outcome, this doesn't reflect on your character.

The cosmologist Carl Sagan, along with almost every astronaut who has seen the earth from outer space, reminds us how important it is to step back and see things from the widest possible angle, poignantly describing our earth as a "pale blue dot."

When seen from so far away, all the boundaries blur and the greatest dramas of history are no more consequential than "mites on a plum." This epiphany is so common among astronauts it has a name: the *overview effect*. In the following thought experiment, I invite you to adopt a cosmic view—one that holds our delusions of self-importance and feeling of primacy with both tenderness and a sense of perspective.

Undoubtedly, a common denominator among most astronauts was the experience of awe. Not surprisingly, Dacher Keltner's lab at UC Berkeley has found a positive correlation between awe and equanimity across seven studies that looked at various factors related to equanimity including temporal distancing, reduced emotional reactivity, and increased physical and psychological well-being due to equanimity. For ideas about how to cultivate awe, read Dacher Keltner's book: *Awe: The New Science of Everyday Wonder and How It Can Transform Your Life*.

PRACTICE

Once again, bring to mind a situation that is troubling you. It could be from an earlier reflection or it could be something new.

Experiment with choosing a situation that feels workable, one in which you have enough wiggle room to entertain alternate perspectives.

Remember the situation in as much detail as you can.

Bring a spirit of scientific investigation to your own experience.

Notice the emotions that are triggered.

Notice the feelings in the body.

Now imagine that you could look back on this same situation in one year.

How might it look?

How would it look to you five years from now?

Ten years?

What if you were in outer space? How would this situation look from the moon?

Can you find it on the pale blue dot?

How does this situation look from the perspective of the universe?

If you find it helpful, take a moment to write your reflections.

The astronaut Ron Garan used an analogy from filmmaking to describe how we can benefit from perspective-taking without using it as a way to ignore or minimize the details and textures of our life circumstances. He talked about a "dolly zoom," where the camera is rolled back at the same rate as the lens is zoomed in. This means, he says, that "the foreground stays the same and the background stretches." (Picture seeing the shark from *Jaws* in close-up as the camera also pans out to capture more of the ocean and the beach.) To "dolly zoom" a situation involves keeping focus on the "worm's-eye details on the ground while zooming out to the longest timeframe possible, ideally multi-generational, while not losing sight of the short term. What I try and do is to live a constant dolly-zoomed life."

Although these are simple reflections you can do in a few minutes and represent teachings echoed throughout most religious and philosophical traditions, they are radically subversive and your mind is likely to rebel! When confronted with reflections that challenge some of our deepest held beliefs about who we are and how best to function in the world, the mind can be expert at spinning narratives about the disasters that will befall us if we let go of certain worldviews.

For example, you may think, *If I fail to take things personally, I will lose my edge. I might become a sap, on the one hand—and just let everyone get away with everything—or I might miss important feedback that could make me a better person.* For me, the fear was being taken advantage of.

How about trying it on and seeing for yourself? Take a day, or a week maybe, and experiment with taking things less personally and see how it goes. What actually happens? Do you lose anything? Do you gain anything? What's really true?

You might worry that if you remember the truth of impermanence, you won't be as motivated, or maybe you won't be able to fully enjoy sensual pleasures, or you'll fall into indifference. Check it out. Take a day or a few days and carry the truth of impermanence into moments of both pleasure *and* pain. Does the truth of impermanence diminish your pleasure? Does it increase your pain?

Personally, I was afraid that, without my agitation, worry, and planning, the sky would fall. I had convinced myself that I needed restlessness and dissatisfaction to motivate me and that my constant planning and busy-ness were holding up the sky. For years I had a day planner (pre-internet, old-school). My planner was very organized! In it, I had lined pages for to-do lists. I was pretty sure my life would unravel without it. At some point I noticed that the to-do lists were actually running me. I lived for crossing things off the list. So I experimented with giving up my precious lists altogether. Cold turkey for about a week. I haven't had a to-do list since (save the occasional complex event or trip).

Perspective-taking can be a powerful tool for bringing ease and balance into life. However, like any powerful tool, its effectiveness depends on how it's applied and under what conditions. Ultimately, you are the judge of what's most beneficial for you. It works best when it's an inside job.

A BOX OF CAVEATS ABOUT PERSPECTIVE-TAKING

Cognitive perspective-taking, as it's called in psychology, can be used as either a powerful tool for freedom and equanimity, or as a cudgel to shame, minimize, or discount suffering. Although you might feel inspired reading Masahide's haiku that opened this chapter, it would be quite a bit different if a friend said to you after learning your house burned down, "Hey, stop complaining; you now have a great view of the moon." You'd probably want to strangle them.

Beware Before You Compare

I noticed during the COVID pandemic that when I compared my own suffering to those I would read about in the news, I would minimize my own pain and feel ashamed of my grief. I didn't believe I had the right to feel badly about my circumstances even though I had "learned" from cancer patients in my groups that you can report without complaining, and it's never helpful to compare our suffering to the suffering of others. When perspective-taking is used as a weapon to force ourselves or someone else to "just get over it"—rather than as a doorway to liberate emotions—it ends up suppressing them.

I had a friend many years ago who suffered one of the worst losses any of us can imagine: her adult son committed suicide, leaving behind a young daughter. This girl, my friend's granddaughter, spent a lot of time with her grandparents after her father died. She loved hiding in a sideboard, where she just barely fit. One day my friend found her crying uncontrollably because she could no longer fit in the cupboard. It would have been tempting to inflict perspective-taking on her: "This means you're growing! Isn't that wonderful?" Instead, my friend understood that, in that moment, her granddaughter's pain was as big as any she had ever felt and simply offered her compassion. Compassion and equanimity offer space for suffering to arise, be known, and pass. Perspective-taking needs to create more space, not minimize or discount.

You Can't Bypass the Pain

While cultivating equanimity can help us experience life with greater spaciousness and ease, we can also easily fall into various forms of spiritual bypass: using the guise of equanimity to dodge the pain. Sorry, it just doesn't work that way. The point here is to make more room for the pain, not to make the pain go away. This is the non-negotiable rub: There's no getting around the pain. Changing our relationship to the pain, making more room, putting it in perspective, does change our experience of the pain and creates room for other feelings. It also gives us the sense of agency that arises from seeing more clearly what we can and cannot do.

Avoid Faking Your Feelings to Defend Against Real Feeling

Another subtle but important distinction is between the posture of equanimity and the genuine expression of equanimity. Both the twelve-step programs and psychologist Paul Ekman agree that "faking" an emotion can be a route to finding the real thing: "Fake it till you make it." Ekman says that assuming the posture and facial expressions of an emotion can be one route to evoking that emotion. At the same time, assuming a posture of what you imagine equanimity should look like can be just another sneaky way of suppressing your emotions.

I spent many years training clinical interns to work with cancer patients. We had an expression we called "sad eyes," when the interns would put on a face to appear compassionate. As you can imagine, it was often affected and yielded the opposite result from what they were looking for. Rather than connecting them with the patients, it alienated them. The posture became more an expression of their own need to appear a certain way, and a defense against their own feelings, than a genuine expression of care and concern.

Intention is what makes the difference. Are you putting on the face and the body of a particular emotion or mental state to create the causes and conditions for that feeling to arise, or are you using the posture as a defense against the feeling? The same behavior, accompanied by very different (often unconscious) intentions, will have very different outcomes.

Chapter 7

WEATHERING STORMS

> The birds have vanished down the sky.
> Now the last cloud drains away.
> We sit together, the mountain and me,
> Until only the mountain remains.
>
> —Li Po

The mountain meditation was made famous by Jon Kabat-Zinn, first through MBSR and later in his book *Wherever You Go, There You Are*. It has been adapted by many people, so you may already be familiar with it. Its popularity may be due to the power and simplicity of evoking a somatic sense of an inner mountain in order to tap into deep wells of equanimity within.

The metaphor of the mountain as a teaching tool is not limited to Buddhist teachings. Cultures across time and space have imbued the mountain with archetypal significance: representing awe, sacred strength, protector, ally, and majesty, to name just a few.

The following guided meditation involves a visualization that allows us to harness the quiet strength of the mountain within our own bodies. This is *almost* a "bottom-up" method in that a template for equanimous responding can be somatically sensed as we align our own bodies with the body of a mountain. This inner geography, once discovered, can offer a map or a pathway back to equanimity when the worldly winds threaten to knock us off-balance.

When working with a guided visualization, it's helpful to allow the process to unfold, and refrain (as best you can) from editing too much. When given a cue to imagine something, go with whatever arises and don't worry about choosing the perfect image. Some people get clear mental images, and other people get more of a felt sense. The power of guided visualizations comes from accessing wisdom that lies beneath or beyond the thinking mind with its tendency toward self-limiting beliefs, closed loops, and judgments.

The following guided practice is adapted from Jon Kabat-Zinn.

PRACTICE

Begin this practice by finding a posture in which you can sit comfortably erect, with a sense of dignity and ease.

Gently close your eyes and then check in with the body, feeling the places where it makes contact with the chair or the floor.

Now take three luxurious deep breaths, completely filling the torso with air, then completely releasing the breath.

Picture a beautiful mountain. Perhaps it's a mountain you already know, or it may be one you create in your imagination. Allow yourself to see its shape, from base to peak, noticing its massive size and solid, unmoving nature.

Sit and breathe with this image for a few moments.

As you're ready, see if you can embody this mountain, sitting from the mountain's perspective—as if you are the mountain.

Experience the mountain's solid base in the lower half of your body, its sides in your shoulders and arms, and its lofty peak in your head.

Feel its elevated nature in the vertical axis of your spine.

Experiencing yourself as the mountain, noticing its centeredness and rootedness in the ground.

Mountains experience all sorts of changes, sometimes on a minute-to-minute basis. Light, colors, and shadows transform the mountain's appearance throughout the day. Wind, rain, lightning, snow, hail, and all forms of clouds arise around the mountain and touch its surfaces.

Throughout all these changes, the mountain just sits, unmoved by the turbulence around it. It remains rooted deep in the earth's crust, stillness in the face of change.

As you sit, see if you can experience this sense of unwavering rootedness.

Regardless of the "weather" that arises, remaining stable and strong, just like the mountain.

Thoughts, feelings, worries, and physical sensations arise and change constantly, much like the weather on the mountain, yet it's possible to remain grounded throughout it all.

Notice, too, the impersonal nature of the weather, and see if you can discover this capacity within, to be slowly "carved" by the weather of life without being disturbed by it.

Imagine the changing seasons from the perspective of the mountain and see if you can adopt this perspective toward the seasons of your own life.

At the age of seventy-two, I have experienced a fair amount of weathering and find it easy to identify with the mountain, in spite of the hundreds of millions of years of difference in our ages. I aspire to be as unapologetic as the mountain and as unconcerned by the world of appearances. Much like the thought experiments in chapter 6, embodying the mountain can help me to find a wider perspective, to take things less personally, and to abide in equanimity.

Chapter 8

JUST LIKE ME

> The first practice is the practice of undiscriminating virtue: Take care of those who are deserving; also, and equally, take care of those who are not.
> When you extend your virtue in all directions without discriminating, your feet are firmly planted on the path that returns to the Tao.
>
> **—Lao Tzu (from the Hua Hu Ching teachings)**

I was thirty years old when I first discovered I was Jewish. I had been confirmed in a Protestant church; raised in Fairfield County, Connecticut; and my birth name was Margaret Anne Nelson. My sister, Elizabeth, and I represented the ultimate assimilation fantasy for our mother, with given names borrowed from the royal family no less. Although a lot of things suddenly made sense (like growing up eating latkes and matzo brei), I wasn't immediately thrilled with this news. I was ashamed as I began to discover subtle threads of anti-Semitism that lurked within me. In a way that would be hard to manufacture, some of the deeply hidden ways that I enjoyed my WASP status were unmasked. However controversial the idea may have become in recent years, unpacking my own white privilege succeeded in powerfully challenging the delusion I held fast to about my own lack of prejudice.

Seeing such biases in myself was important on many levels. The most obvious is that what is not acknowledged cannot be remedied. But on a deeper level, it put me squarely on a continuum with everyone who harbors prejudice against any group anywhere. I could no longer place myself above extremists who believed their race to be superior to another race because then I would just be "othering" that group. The shame I felt helped me to understand why so many people are in denial about their own bigotry and how blinding that denial can become. It was through discovering my own shadow, the yuckiness of the shame I felt, and the deep pernicious roots that nourished it, that I could begin to feel in the marrow of my bones that "just like me," everyone else was doing their best to be a person. It's hard to be a person.

Out-group bias often begins in subtle and unremarkable ways. It can start through simple preferences that begin innocently enough: I prefer mayonnaise to butter on my sandwiches, and I like blue green better than olive green. In mysterious and unconscious ways, these preferences start spilling over to people. From there they become entrenched through media portrayals, as well as through cultural, familial, and institutional structures and habit patterns.

Here again, I turn to a favorite children's book for insight and wisdom. Rosemary Wells wrote a charming series of books about Yoko, a Japanese kitty who was probably about six years old. One day, Yoko's mom packs sushi in her lunch box and the other kids find this utterly distasteful: "Ick! It's seaweed. Yuck-o-rama." Her kindly kindergarten teacher, Mrs. Jenkins, assesses the problem and comes up with a brilliant solution. She sends a note to the parents announcing international food day at the school. Each child should bring a dish from a foreign country and every student must try a bite of everything.

Though the class is slow to take to the idea of sushi, eventually one boy is brave enough and hungry enough to dare take a bite and proceeds to gobble up the whole platter.

It would be wonderful if kindergarten teachers the world over invited young minds to catch their prejudices early, before they hardened into bias, bigotry, and a kind of tribalism or elitism capable of massacring untold numbers of innocent people. As Hannah Arendt confirmed in her seminal two-part *New Yorker* article of 1963, the incomprehensible evil of Adolf Eichmann was rooted not in psychopathy but in something much scarier and more insidious. He was simply a mundane, ambitious, rule-following bureaucrat who was exposed to cultural, political, and personal influences that allowed him to progress on the continuum of prejudice from innocent preference to wholesale massacre.

This terrifying possibility was also depicted cinematically in the chillingly beautiful 2023 feature film *The Zone of Interest*, based on the Martin Amis novel of the same name. Here again, a high Nazi official, the commandant of Auschwitz, lives a happy "normal" life, as do his wife and five beautiful children in a gorgeous villa abutting the concentration camp. Though the story is only loosely based on the historical family, we see a plausible yet mind-boggling portrait of a man able to love his wife and play with his children while at the same time innovating the most effective and systematic way to organize mass murder. Thanks to Commandant Rudolf Höss, Auschwitz's gas chambers were capable of slaughtering two thousand people an hour.

There are many credible theories positing that out-group and in-group biases evolved for reasons of survival and are therefore "hardwired" into our nervous systems. This might be one of the many explanations for how easy it is to activate them and for our seemingly inexhaustible capacity as human beings to dehumanize out-groups and justify all kinds of violence, up to and including warfare.

Non-Bias as a Mark of Equanimity

Impartiality is a hallmark of equanimity and is central to the Buddhist teachings, particularly following the period when Mahayana Buddhism placed the bodhisattva vow front and center on the Buddhist path. The bodhisattva vow is the commitment to relieve the suffering of all beings without exception. Just as with mindfulness practice, Buddhism offers a vast and highly developed system for training the heart in impartiality. In fact, there's a classic practice in Buddhism specifically designed to overcome bias (a version of which you will find on page 137). Once again, don't be fooled by how simple it might seem. It is both profound and has the seeds to transform society, heart by heart. Like most Buddhist practices, it came from looking deeply at the nature of mind and seeing how it was that bias took root and became toxic, and what were the most powerful antidotes. The extent to which the Buddha understood human psychology 2,600 years ago never ceases to amaze me. It seems that bias is a perennial problem and it takes a sophisticated understanding of its root causes to uproot it.

As I write, the world's attention has shifted from the horrors in the Ukraine to the bloodshed in Gaza. University campuses are erupting in protest against the brutality of this war. October 7, 2023, blew the lid off a long-standing conflict and the whole world has been forced to grapple with the tragic and heartbreaking consequences of armed conflict over land, resources, religion, and culture.

The outbreak of this war broke my heart. I imagine it broke your heart, too. Where do we start with problems like this that are so entrenched and seemingly intractable, so beyond our realm of influence? I doubt that any of us can draft a two-state solution or facilitate diplomatic negotiations. So where *can* we begin? The worst, and yet most common response, is to add our own outrage-, anger-, and adrenaline-driven blindness to a situation already riddled with blindness. And yet our anger is critical in signaling the injustice we feel at mass killing that is aimed at fostering peace and security.

What then? Is the only alternative to throw up our hands in despair? However tempting that may feel, that may be the second-worst option. Despair is one of the near enemies of compassion. Despair takes you out of the game. Compassion keeps you in the game. And understanding our interconnection is a gateway to compassion.

Zen master Thich Nhat Hanh wrote extensively about the idea of *interbeing* and how important it is to recognize the inexorable interconnection of everything. Looking beyond the surface, it's easy to see how nothing exists independently and we are all utterly reliant on a myriad of beings and elements for our basic survival, let alone for our well-being.

In a YouTube video long before October 7, 2023, he said:

Look at my two arms. My right arm can do calligraphy. It has written hundreds of poems. It can ring the bell. And yet my right hand is never, never proud of itself. My right hand will never tell my left hand, *You, you're good for nothing. You don't write any poems. You don't practice calligraphy.* Why? Because my right hand has the wisdom of equanimity. We call this the wisdom of nondiscrimination.

The Israelis and the Palestinians should look at each other like the right and the left hands. The wisdom of nondiscrimination can bring about true peace, true love, and it'll help to remove the fear. I don't think that the right hand is afraid of the left hand. And the left hand isn't afraid of the right hand because both of them have the wisdom of nondiscrimination. They know that they belong to each other. They are inside of each other, and everything that happens to the right hand will happen to the left hand. The suffering of one finger is the concern of all fingers, all ten fingers.

Othering and bias are at the very root of so much suffering in the world. Though there has been tribalism throughout history, most of us in the US have never before lived under a demagogue who thrived on fear and tribal hate. Eschewing tribalism doesn't mean that you can't have pride in your heritage.

So what can we do about a problem like Gaza? We can start by reducing our sense of othering and tribalism and superiority. Generational divide in Jewish families became especially poignant when the Gaza crisis erupted, just as it had been over Trump's foreign policy. Civil wars have broken out even within families. Jewish students arrested for protesting against Israel have parents and grandparents who remember the Holocaust and are saying to them, "What's the matter with you?"

May we all step back from the brink and find our shared humanity—even within our own four walls. This practice may help.

"Just Like Me" Practice

Before starting, remember that although true feelings of compassion and equanimity can't be forced, these practices tenderize the heart whether or not they "feel good" in the moment. Although the focus of this practice is to nourish compassion and overcome bias, other feelings will also arise, and this is both normal and important.

If sadness arises, it can be embraced as part of a grief process and an indication of how the heart heals itself. If anger arises, see if you can hold it with tenderness, not forcing anything, while also giving yourself permission to focus on the breath or a neutral sensation in the body.

The following guided practice includes six different expressions of "Just like me." Feel free to use them all, to write your own, or to choose the ones you prefer. Should you choose to continue working with this practice, you can streamline it by choosing one or two phrases that become placeholders for the depth and scope of feelings the phrases are meant to evoke.

PRACTICE

Begin by experimenting with a posture that lets your back be strong and your front soft. Settle into this physical metaphor for equanimity. Allow the posture itself to set an intention to be open-hearted and courageous.

Now bringing your attention to the area around your physical heart—noticing how this area gently expands or fills when you breathe in; and how it relaxes when you breathe out.

For a few breaths, imagine you could inhale directly into the center of your chest, expanding the chest, lungs, and heart. Imagine you could exhale directly from the center of your chest.

As you gather your attention on the sensations of the breath in the chest, bring extra gentleness to any tender feelings you might become aware of.

As best you can, take the next few minutes to direct your attention to the sensations around the heart as you breathe.

Whenever you notice the mind has wandered to thoughts, images, sounds, or sensations in other parts of the body, gently escort it back to the sensations in the chest as it rises with the in-breath and falls with the out-breath.

Now, bring to mind someone you care about, such as a family member, mentor, or friend, someone who naturally brings a smile to your face.

Try to vividly feel their presence in front of you.

Notice how you feel when you think about this person.

Reflect on the complexity of this person's life.

Just like you, they have felt joy and sorrow.

Just like you, they have a body that feels both pleasure and pain.

Just like you, their heart has been broken.

Just like you, this person has goals and dreams.

Just like you, this person is an object of deep concern to someone; they are a child to someone; a parent, or spouse, or dear friend to someone.

Just like you, this person wants to love and be loved; to contribute, and to be appreciated.

With this in mind, repeating silently, "Just like me, this person wishes to be happy and free from suffering."

See if you can feel into the fundamental truth of this statement. "Just like me, this person wishes to be happy and free from suffering."

Release the focus on a loved one and bring your attention back to the heart center. Feeling the rise and fall of the chest, as well as any feelings of tenderness that may arise.

Now bring to mind someone you would recognize but have no special sense of either closeness or conflict with—a stranger. It could be a person you see at the grocery store or coffee shop; someone you work with but don't know well; or perhaps a neighbor you see around but don't know.

Try to vividly feel their presence in front of you.

Notice how you feel when you think about this person.

Imagine the complexity of this person's life.

Just like you, they have felt joy and sorrow.

Just like you, they have a body that feels both pleasure and pain.

Just like you, their heart has been broken.

Just like you, this person has goals and dreams.

Just like you, this person is an object of deep concern to someone; they are a child to someone; a parent, or spouse, or dear friend to someone.

Just like you, this person wants to love and be loved; to contribute, and to be appreciated.

With this in mind, repeating silently, "Just like me, this person wishes to be happy and free from suffering."

See if you can feel into the fundamental truth of this statement. "Just like me, this person wishes to be happy and free from suffering."

As you release the focus on a stranger, notice if you still feel "neutral" toward them. Perhaps considering their humanity in this way has brought feelings of warmth in the heart.

Now bringing your attention back to the heart center and the simple sensations of the chest rising and falling as you breathe.

Now bring to mind a person you have some difficulty or discomfort with. Maybe it's someone you don't get along with, or feel in competition with. Maybe it's someone you think has caused you harm or someone who doesn't seem to share your beliefs or values.

As best you can, vividly feel their presence in front of you.

Notice how you feel when you think about this person.

Imagine the complexity of this person's life.

Just like you, they have felt joy and sorrow.

Just like you, they have a body that feels both pleasure and pain.

Just like you, their heart has been broken.

Just like you, this person has goals and dreams.

Just like you, this person is an object of deep concern to someone; they are a child to someone; a parent, or spouse, or dear friend to someone.

Just like you, this person wants to love and be loved; to contribute, and to be appreciated.

With this in mind, repeating silently, "Just like me, this person wishes to be happy and free from suffering."

See if you can find any truth in the statement: "Just like me, this person wishes to be happy and free from suffering."

Remember that it isn't necessary to love them or even to forgive them, and it is never helpful to insist that you feel any particular way.

Just see if you are able to recognize your shared common humanity and see what happens from there.

Bringing your attention back to the heart center and the simple sensations of the chest rising and falling as you breathe, bringing respect and tenderness to what you have asked of your own heart.

Now take a moment to picture all three people together in front of you.

Remembering how they all have complex lives, hopes, and dreams. How we all want to be happy and avoid suffering. On this level, there is no difference between any of us. In this fundamental respect, we are all the same.

Now, continue to expand the scope of your awareness to all those around you—imagining that your heart is like an energy field that can expand in all directions—wider and wider—to include everyone who lives in your state, on your continent, and on this planet.

Letting feelings of care and kindness extend to those who are suffering and to those who are happy.

To those who are old, and perhaps dying in this very moment...

To those just being born, with their whole lives ahead of them.

To those with great wealth, and to those barely scraping by.

To those in your political party and to those of other political parties.

To those of your tribe, to those of other tribes.

To the oppressors and to the oppressed.

Appreciating this great mysterious web of life that connects us all and the fundamental aspirations we share with all beings.

Again, bringing your attention back to the heart center and the simple sensations of the chest rising and falling as you breathe.

Should you choose to work with this practice, a variety of feelings will come up. Feelings of anger, sadness, anxiety, and fear are just as important to surface as unconditional goodwill. They may be just the feelings that are blocking the generous outflow of love. Just as Rumi said in his famous poem "The Guest House," welcome and entertain all your feelings because you never know when they "may be clearing you out for some new delight." Celebrate the capacity to increase your "window of tolerance" and remember how this, too, is linked to reducing bias.

These practices are meant to be done over and over again. Whether you're brand-new to this practice or you've done it hundreds of times, I encourage you to keep at it. We are all so complicated and messy. We are the products of countless interactions that have configured our hearts with mysterious channels, barriers, and cul-de-sacs.

Really Showing Up

In chapter 16, we will learn from climate scientist and Zen Priest Kritee Kanko about how she teaches people to harness grief and anger in the service of wise action in the world. Kritee helps us engage with a classic practice of cultivating an understanding of shared common humanity—the essence of non-bias—without blurring over or minimizing important distinctions that acknowledge the massive differences in our circumstances and histories. Though we share many fundamental experiences with all sentient beings, particularly the human variety, we are not all in the same "boat." This became achingly clear during the COVID pandemic. While everyone across the globe was confronting a life-threatening virus, some of us were in "yachts" with access to the latest medical care while others of us were in rafts with minimal access to the basic necessities for survival.

When I asked Kritee how to reconcile this practice of shared common humanity with the vast inequalities in the world, she shared one of many key distinctions with me. Rather than the term *inequality*, Kritee prefers the phrase *different positionality*. This subtle shift of words revealed a universe of habit and desensitization to me. Of course we are not unequal. We are all equal; however, we find ourselves in vastly different positions politically, educationally, economically, and socially due to complex causes and conditions. Kritee suggested that, without this recognition in our bones, and embodied in our lived experience, we aren't really in a position to talk about shared common humanity:

> If I have done the work of standing in solidarity with the communities and people I'm talking to, if they know I would give up my time, money, resources to help them, then they might be open to hearing me express my understanding of shared

humanity. Otherwise, going on about shared humanity is bullshit. The words mean nothing. Humanity means my pain is also your pain. Are you really going to take compassionate action or are you just talking about compassion? Are you going to stand with me when police unjustly arrest my son because someone else's house was broken into? You have to ask yourself: Are you showing up with your body or are you just showing up with your words?

Imagine

Sometimes after I do this practice, I think of the great John Lennon song "Imagine." It's a balm to my heart to imagine "all the people . . . living life in peace." I recently heard a presentation by Israeli academic Eran Halperin about extensive polling that his group, aChord, had been conducting in Israel weekly since October 7, 2023. Not surprisingly, Israeli's number one concern in that poll was security. As I do this practice of extending care across conceptual categories, I truly believe that most people want to live in peace. I believe that most Israelis, Palestinians, Russians, and Ukrainians want their children to be safe. I believe we share many fundamental values and concerns with one another.

Popular news sources derive revenue from activating intense emotions, and social media algorithms reward anger, fear, and anxiety. In that way they fuel and exacerbate othering and demonizing. As such, maintaining faith in humanity can feel like swimming upstream. There may be a voice in your head chiding you for being naive, Pollyanna, overly idealistic, unrealistic, cheesy. That kind of cynicism is a defense against fear. If I don't expect anything, I can't be disappointed. By contrast, the whole project of equanimity is a willingness to live fully with a broken heart.

Practicing impartiality is an important step in overcoming bias. It's an inside job and we can only do this for ourselves. In chapter 16, we'll talk more about bringing equanimity into activism, but it all starts in the crucible of our own hearts. Celebrate the practice of overcoming bias. Every time, every minute we overcome bias in our own minds we contribute to the world we want to inhabit. This might seem like a very small thing, but it's more powerful than you might imagine, and you can start right here, right now. Our upside-down world needs every single drop of equanimity we can offer.

Chapter 9

HOW TO LOVE AND CARE WITHOUT ATTACHMENT

There is only the trying. The rest is not our business.

—T. S. Eliot

In chapter 1 we explored how understanding something intellectually can trick us into believing we have internalized a principle sufficiently to have it shape our behavior in the world. The thought experiments in chapter 6 were designed to challenge how deeply we rely on core beliefs about the nature of time, space, and self when we engage with the world, particularly when things are uncomfortable. Probably most of you would agree, in theory, that we are each responsible for our own happiness. However, this intellectual understanding is frequently overridden when it comes to those closest to us, resulting in untold layers of extra suffering.

There's a beautiful metaphor from the Buddha that describes this kind of suffering as the "second arrow." The first arrow that strikes us represents the unavoidable pain that comes with life. The second arrow represents the mental anguish, resistance, or aversion we create *in response to that pain*. In effect, we shoot ourselves with that second arrow. Most of our second arrows derive from misunderstandings about the nature of reality. We take personally what is impersonal, forget the truth of impermanence, expect the world to always be pleasant, or believe that we have the god-like power to prevent those we love from suffering—which is what we will focus on in this chapter.

Certain relationships can make it particularly challenging to remember the limits of our influence on the happiness of others. For me, this has been most apparent in my role as a mother. It is both sobering and illuminating to acknowledge that the insight I had all those years ago in the desert—that it was my job to save my mother from depression—no matter how profound, is no match for the conditioning and biology that compel me to take responsibility for my child's happiness (even when she's thirty-one years old). Deeply conditioned beliefs such as these require patience and constant reminders.

Though we may differ in our views regarding free will, divine intervention, and determinism, we probably agree that we all exist in a complex web of interconnection. It's hard to point to any single cause for an individual's sense of well-being, happiness, or discontent. There are circumstantial factors, historical factors, biological, physical, emotional, psychological, and spiritual factors, contextual and geographic factors, political factors, and on and on.

As parents, of course, we understand this. We know that even newborns suffer before we had much influence over their lives. My daughter was in the NICU, so I saw that firsthand. And yet, for those close to us—especially our children, our siblings, our parents, our close friends—we take on a level of responsibility for their happiness and well-being that goes far beyond our ability to control. We may even become rescuers, imagining that if we just care hard enough and work hard enough, to the point of burnout, we can control the outcomes for those we're attached to.

As my daughter faced inevitable challenges, I strove mightily to protect her from suffering, and probably managed instead to offer a poor model for learning how to tolerate distress. I'm not in a good position, then, to tell *you* how to do that, but I have learned that I cannot control others' circumstances and the consequences of their previous actions.

Another important teacher of mine was Marshall Rosenberg, founder of the groundbreaking Nonviolent Communication methodology. I'll never forget Marshall saying,

> You can't make your kids do anything. All you can do is make them wish they had. And then, they will make you wish you hadn't made them wish they had.

There is no one-size-fits-all formula to respond to the suffering of others. Each family, each circle of friends, each community, will bring their own temperaments, conditioning, ancestors, cultures, circumstances—the full tapestry. Therefore, we find the way forward embedded in the reality of those circumstances.

I learned of a different—and seemingly more equanimous way—of relating to those close to us, especially our children, from a conversation with Kiliii Yüyan, the photographer we met in chapter 2, who has chronicled the lives of Indigenous people in some of the most remote corners of the world. Kiliii told me,

> Native peoples don't really teach things. As younger people, we simply watch what older people are doing and we do those things. There's a little bit of mentorship that happens naturally, but by and large it's a watch and explore and do and try again.

In this way of approaching things, there's less attempt to control the outcome and more sense of letting the other person rely on themself. It's no less caring, but it is less cloying, and it avoids the illusion that we can control others' circumstances. My husband and I met an emergency room doctor with thirty years' experience at a party, and Michael asked her with genuine curiosity how it was possible to tolerate the constant state of emergency. She replied:

It's not my emergency.

That's become a catchphrase for us. It helps us see how easily drama can hook us, and cause us to get sucked into the maelstrom and lose any effectiveness we might have had—because our efforts to be helpful are laden with our own stress. The person hoping we can rescue them in one fell swoop may say something like, "If you really loved me, then you would . . ." They believe that if you don't jump into the drama, you don't care. But if you do jump in, you can make it worse.

When someone is in desperate pain and fear, they may imagine we can simply fix it. We are thereby invited to jump into a shared delusion. When any of us is in a critical amount of pain, we may ask for a lifeline that can't be provided. When it doesn't work, we may resort to a barrage of words to demonstrate our caring, and start pitching a string of possibilities and solutions, yet more and more words may simply make it worse. There's a poignant sense of helplessness and we all want to avoid feeling it, accepting it.

To bring equanimity to bear when we're asked to help beyond our capability, we need both wisdom and compassion—two wings of a bird. Wisdom allows us to discover that there isn't a compassion-er who will solve it all. The compassion is simply the warmth and caring that can radiate even when we can't control the outcome.

In this chapter, as a method of cultivating equanimity that respects our inability to control others' circumstances and outcomes, I would like to share with you a classic Theravada Buddhist practice that gave rise to the liberating insight I had about my mother that I recounted in chapter 1. The simple yet profound phrases in this meditation have likewise impacted hundreds of people in my classes, let alone all the meditation practitioners who have worked with this practice over millennia.

For the following practice I encourage you to experiment with a simple posture that I learned from Roshi Joan Halifax (originally offered by Trungpa Rinpoche) and briefly mentioned in the last chapter. I've been using it and teaching it for years in my meditation classes. Although it is helpful for meditation in general, it's particularly useful in the cultivation of equanimity. She invites us to take our seat with a strong back and a soft front. A soft front suggests the heart is open and vulnerable, able to be touched by and respond to the world. A strong back gives us the courage to meet what is painful and scary and to stay upright, like the gyroscope, as the worldly winds attempt to blow us off course.

Consider, too, the opposite of a strong back and soft front. We've seen examples in political leaders of a puffed up and defended front incapable of empathy and a soft back that doesn't have the strength to maintain a moral course in the face of threats to self-interest. Who would you like to be?

In the following practice, we'll experiment with several different sets of phrases that are reminders of wider views and deeper truths that can be hard to access from everyday habituated mind states and mental patterns.

Once again, mental states like equanimity and compassion can't be forced. This practice is a way of creating the conditions for equanimity to arise. If you encounter resistance, it may be that there are feelings that need to be recognized, felt, and acknowledged before it's appropriate or helpful to bring in the wider perspective of equanimity. Should that be the case, I encourage you to either choose a less triggering situation or shift to a practice of self-compassion for as long as necessary before coming back to equanimity practice.

The following words have been drawn from my own teaching as well as guided meditations from several of my teachers, most notably from Jack Kornfield, who generously allowed me to share them with you in this book. Although the language is carefully chosen to be as universal as possible, should you find yourself getting derailed by the wording, please feel free to change it to suit your own conditions and frame of reference.

PRACTICE

Beginning with three deep diaphragmatic breaths... filling the entire torso with the in-breath, from the belly all the way up to the collar bones, and emptying the torso completely on the out-breath. Like filling and emptying a vessel of water.

After the third out-breath, release any control of the breath and allow it to find its own natural rhythm. As the breath settles into a more natural pace, gently gather your attention on the sensations of the breath, wherever they're most predominant in the body.

As best you can, bring an attitude of kindness, patience, even amusement, to the wandering mind. It's inevitable, so why fight it? And, the less you fight it, the less agitated it will be.

If you're able to stay with the sensations of breath in the body for a breath or two, see if you can notice the space in between the breaths. That moment of perfect equipoise that happens all by itself, on average fifteen times per minute. Let this naturally arising moment of balance in the body become the focus of your attention. See if you can find the effortless, spacious, ease between expansion and contraction, between rising and falling, in and out.

Reflecting for a few minutes on why you've chosen to read a book about equanimity. What are the benefits of a mind that has balance and equanimity?

Perhaps sensing what a gift it would be to bring a peaceful heart to the world around you. Connecting with your own deepest and most sincere intention to cultivate equanimity by silently repeating these simple phrases:

"May I be open, balanced, and at peace."

"May my heart be at rest and balanced in the midst of all things."

Sensing into that point of balance within yourself amidst all that is constantly changing, your inner gyroscope.

Experiment with the following phrases, and see if they're true for you and help you find greater spaciousness and ease:

Things are just as they are.

All things are impermanent.

Feelings arise and pass away.

I am safe in this moment.

As you repeat these phrases, connecting as best you can with their full implications for whatever upsets or preoccupies you in this moment:

Things are just as they are.

All things are impermanent.

Feelings arise and pass away.

I am safe in this moment.

Reflecting for a moment on the reality of how much our own happiness or suffering is a result of our thoughts, actions and circumstances, and not others' wishes for us. We can have partners who love us, friends, parents, but we are ultimately responsible for our own happiness and well-being.

This is also true for our loved ones. Loving others inevitably opens us to feelings of helplessness when our loved ones suffer. And, no matter how fortunate their lives, there will be periods of pain for everyone. Reflecting on the idea that all beings are the recipients of their own choices, actions, and circumstances.

While you can deeply love and care for others, in the end, their happiness and suffering depend more on their thoughts, actions and circumstances, than on your wishes for them.

Equanimity practice recognizes this reality without forsaking any of the love that you feel.

Now bringing to mind someone you love who's going through some difficulty and feeling into the following phrase:

"Your happiness and suffering depend on your thoughts,

actions and circumstances, and not my wishes for you. Recognizing this, I will continue to wish for your happiness, and for you to understand the deepest sources of your own happiness."

"I will care for you but cannot keep you from suffering."

Knowing and accepting the limits of what you can do for this loved one, offer them the following phrases:

"May you be happy and have access to the deepest causes of your own happiness."

"May you be balanced and peaceful."

"May you find true equanimity."

Allowing your heart to be touched by whatever poignancy you experience as you imagine your loved one being free from suffering, while fully recognizing the limits of what you can and cannot do.

Perhaps refreshing the image of your loved one, and silently repeating the same phrases:

"May you be happy and have access to the deepest causes of your own happiness."

"May you be balanced and peaceful."

"May you find true equanimity."

Now imagine expanding your own field of equanimity beyond yourself and your loved one, to encompass the whole world:

"May I bring compassion and equanimity to the events of the world."

"May I find balance and equanimity and peace amidst it all."

To close this practice, release the words and images and return to the simple sensations of the breath in the body, relaxing into the rhythm of expansion and contraction and savoring the natural point of balance between the two.

However you choose to work with this practice, it might be helpful to investigate your beliefs about your responsibility vis à vis the happiness of your loved ones and friends. How do you respond and who is it serving? What ideas are you carrying about how you should behave as a loving parent, partner, or friend? Are you squandering energy on trying to control what you can't control? Does your idea of what a loving "X" (mother, husband, friend) looks like deplete you or create resentment? Many of us are taught that martyrdom is a measure of love, or we confuse love with attachment. Loving without attachment may seem cold and off-putting. Here again, we can be squeezed into a false binary: either we take full responsibility for our loved ones or we must not care.

I'll never forget a story that Marshall Rosenberg shared at one of the many trainings I did with him. Marshall held fast to the radical belief that everyone could get their needs met. He patently rejected the zero-sum idea that love was finite and could only be apportioned: If Sam gets 60 percent, then there's only 40 percent left for Olivia. In this way, he was very much in sync with the Buddhist idea of immeasurable love.

He shared a story about a woman who came to his workshop complaining that she couldn't get her needs met because she hated cooking for her family. Her family needed to eat and her role as mother and wife required her to feed them. Either *they* got their needs met or *she* did. For Marshall, there was a very simple solution: don't cook. In her mind, though, not feeding her family was the same as not loving them. She had trouble imagining that a loving mother could possibly make such a choice.

A few years later, a young man approached Marshall in a workshop. He reported that his mother had announced two years ago that she was no longer cooking for the family. He couldn't thank Marshall enough. In fact, the whole family was so thrilled they all took workshops with him. "Her food was terrible," the son said. "We hated it, but we didn't want to hurt her feelings." Perhaps it was poisoned by her resentment? (There is, in fact, a charming novel, *The Particular Sadness of Lemon Cake*, by Aimee Bender, in which a nine-year-old girl can "taste" the longing and wailing of her mother through the food she prepares.)

Marshall offers a developmental view on emotional responsibility. This involves three stages:

1. **Emotional slavery**—believing ourselves responsible for the feelings of others

2. **The obnoxious stage**—in which we refuse to admit to caring what anyone else feels or needs

3. **Emotional liberation**—in which we accept full responsibility for our own feelings but not the feelings of others, while being aware that we can never meet our own needs at the expense of others

Being liberated from the myth that we control others' feelings is truly a gift to ourselves and to those we love, and yet another gateway into equanimity. At the same time, it's important to remember that the near enemy of equanimity is indifference. There are countless ways we intervene with our loved ones both to prevent suffering and to comfort them. The intention of this practice is to strengthen the wisdom that allows us to understand which things we can control and then let go of those we can't.

Chapter 10

STEPPING STONES TO EQUANIMITY

A day without sunshine is like, you know, night.

—Steve Martin

A short sentence can pack a punch. In chapters 8 and 9, the guided meditations included phrases that were meant to be repeated silently to invoke different flavors of equanimity. Phrases like these function as shorthand placeholders for deeper, more complex wishes that can be both woven into meditation practice and also evoked in everyday life. I have collected dozens of them and share them with you here.

First, a few words about how to work skillfully with these phrases. There are similar exercises—such as silently reciting mantras or intoning affirmations while looking in the mirror—that have become ubiquitous in everyday parlance. Much like the word *mindfulness*, the word *mantra* has become so common that it has lost its original meaning. People often talk about pet phrases of encouragement, praise, or self-compassion as their "mantras," whereas the original function of mantra in the Vedic tradition was evoking the sacred and was deeply connected with religious ritual.

As mantras became appropriated by the self-help movement, they began bleeding into the idea of affirmations. Affirmations are popular with various New Age teachers who tend to convey the same message: If you think positively, repeat phrases frequently, and believe magically, you can be happy and successful and have whatever you want. For example: "Visualize it and it will come to you" or "the law of attraction always works."

This is dangerous on many levels. First, it can be a setup for yet another way to beat yourself up when things don't go the way you would like them to go. "If only I had a better attitude, then fill in the blank wouldn't have happened." Second, it tends to reify our self-importance, which gets us into all sorts of trouble by overinflating our power and disconnecting us from others. From the perspective of equanimity, using affirmations in this way also disconnects us from natural laws and moves us out of balance. Finally, it can actually add to our suffering when it doesn't prove effective.

Some people see suffering as the difference between what we wanted and what we got. What increases that differential? What decreases it? If you expect life to be up and down, you will be more peaceful when it is exactly that. Equanimity is about finding balance amidst the constantly changing circumstances of life rather than insisting that they conform to our preferences.

The great Burmese meditation teacher Sayadaw U Tejaniya likes to say, "You only want good experiences. You don't want even the tiniest unpleasantness. Is this fair?"

By contrast, these phrases are intended to connect you with the truth of life's unfolding with the greatest possible ease and grace. They come from many teachers I've encountered over the years, and I share them with you to read through at your leisure. Some are just subtle rewordings of the same phrase. Feel into them and see if they signal another way of relating to your experience with greater wisdom and spaciousness.

These simple phrases can serve as reminders of fundamental truths. Truths that are easy to forget, especially when we get caught in the grips of social media and the twenty-four-hour news cycle. If you already have a meditation practice, you can drop in any of these phrases in whatever way feels helpful to you. You might choose to shift the language of one of the guided meditations that accompany this book to match a different phrase that speaks to you more personally. You might choose a phrase or two to recall at the start or end of any day.

What I do know is that whatever happens, equanimity can only help. For your own well-being, for those in your immediate circle, and for the world at large, I hope you find multiple doorways into equanimity through these phrases.

Special thanks to Nikki Mirghafori, Christina Feldman, Joseph Goldstein, Jack Kornfield, Gil Fronsdal, Sharon Salzberg, Sylvia Boorstein, and the great equanimity master, Barack Obama.

May I meet this moment fully. May I meet it as a friend.

This is what is happening now; I wonder what will happen next.

Anything can happen at any time.

There's a mystery here. Things take time.

I'm not in charge of the universe.

Things are just as they are right now.

It is what it is right now.

This *is* the curriculum.

Of course.

This is how it is. You can be sad, but outrage is extra.

Ah, this is what [fill in the blank with any possible experience] is like.

May I learn to see the arising and passing of all nature with equanimity and balance.

May I be open, balanced, and peaceful.

May I remain peaceful and let go of expectations.

May I find the inner resources to be able to give to others and receive myself.

May I bring compassion and equanimity to the events of the world.

May I find balance, equanimity, and peace amid it all.

May I see my limits compassionately, just as I view the limitations of others.

May I embrace change with stillness and calmness.

May I deeply accept this moment as it is.

May my home be a home of balance and spaciousness.

May I remain unshaken by life's rise and fall.

May I rest in not knowing.

May I be equally near all things.

May my heart be at ease with the changing conditions of life.

May I accept the arising and passing of all things with equanimity and balance.

No matter how I might wish things to be otherwise, things are as they are right now.

May I accept this just as it is.

May I open to and accept how it is right now because *this is* how it is right now.

Your happiness and suffering depend on your thoughts and actions and not my wishes for you.

May you learn to see the arising and passing of all things with equanimity and balance.

May you meet all your experiences with ease and kindness.

May you have true equanimity.

May you be balanced and peaceful.

I will care for you but cannot keep you from suffering.

May I see you as I wish to be seen: as big as life itself, so much more than your needs or pain.

May I offer love, knowing I cannot control the course of life, suffering, or death.

I care about your pain yet cannot control it.

May I be open to how it is between us right now with more balance and ease.

All beings have their own journey, according to unfolding conditions.

I honor your life's journey.

I care for you, but I can't control your happiness and unhappiness.

I care about you, but I'm not in control of the unfolding of events. I can't make it better for you.

All beings are owners of their actions, heirs to their actions. Their happiness and unhappiness depend on their actions, not on my wishes for them.

All beings have their own journey.

Whether I understand it or not, things are unfolding according to a lawful nature.

I wish you happiness but cannot make your choices for you.

All beings meet their joys and sorrows according to a lawful nature.

May we all accept things as they are.

May we be undisturbed by the comings and goings of events.

The only thing that's the end of the world is the end of the world. (Barack Obama's message to his daughters from his final press conference before leaving office and right after the election results.)

Chapter 11

BOTTOM-UP EQUANIMITY

> Mr. Duffy lived a short distance from his body.
>
> —James Joyce

In 1637, René Descartes gave us the famous line "I think, therefore I am" in his *Discourse on Method*. Alongside this formula, he also asserted a divide between mind and body. While finding the origin of consciousness in a world of matter remains the "hard problem" for science, we're starting to grasp the two-way street between our brains and bodies. Intelligence, once thought to be exclusively in the head, now shows up in surprising places, inside and outside of us.

Consider the brain-gut axis: William Beaumont hinted at it in the 1840s, but it took until 1999 for Michael Gershon—a neurogastroenterologist (whew!; try saying that at a party)—to explore it fully in his book *The Second Brain*. This "little brain" in our guts not only affects our overall health but also our mood.

Several phenomena have kept us from fully appreciating the body's wisdom. Descartes's old ideas about the mind-body split, our love for rationality, and a focus on fixing what goes wrong rather than understanding the body's natural intelligence—all of these have played a part. Western culture tends to praise brainpower while overlooking the wisdom in bodily experiences and intuition. Plus, Freud's theories didn't help, either, painting the body as a battleground between the unruly impulses of the id and the high moral standards of the superego.

It's no surprise that therapeutic interventions have traditionally favored top-down approaches, where cognitive processes serve as both the entry point and the focus of attention. However, bottom-up approaches like somatic experiencing, EMDR (eye-movement desensitization and reprocessing), hakomi, bioenergetics, and other modalities have been gaining traction in recent decades, especially as we deepen our understanding of trauma and its most effective treatments. When Westerners adopted Eastern meditation practices, they also tended to emphasize the mind over the body, despite the fact these practices in their original contexts had a strong bodily component and were practiced by people who did not live sedentary, technologically assisted lives.

Of course, bottom-up and top-down are not entirely discrete. They exist in a network. Most mindfulness practices, properly practiced, involve the body, and it's impossible to completely exclude the mind from any bottom-up approach. The dichotomy between top-down and bottom-up approaches is inherently false. Our evolving science is bringing us back to wisdom that Eastern traditions have understood for millennia: heart, body, and mind are one. In languages such as Chinese, Tibetan, Pāli, Sanskrit, and Japanese, single words encompass both heart and mind.

However faulty the mind/body dichotomy might be, because we have favored so-called top-down approaches, many of us need to explore bottom-up modalities in order to come into balance, let alone as a more skillful way to work with trauma.

What might it look like to cultivate equanimity in a bottom-up approach? Is it possible? Might it be better in some cases?

Sense Foraging

This natural evolution from top-down to bottom-up to "heartmind" can also be traced in the career arc of Zindel Segal, a Canadian clinical psychologist who co-developed Mindfulness-Based Cognitive Therapy (MBCT). I first met Zindel at an early teacher training for MBCT in Southern California and was deeply impressed by his warmth and clinical aptitude. It made perfect sense to me that mindfulness was the "secret sauce" that cognitive behavioral therapy (CBT) needed to boost its efficacy—especially in the context of relapse prevention for depression, which was Zindel's passionate aim. Mindfulness provided both the techniques and the attitudinal foundations that made it possible to relate to thoughts differently—with more spaciousness and less identification—foundational to CBT.

As the name suggests, CBT began as mainly a top-down approach focusing on distorted or unhelpful thinking as the primary intervention. I remember a conversation with Zindel many years ago in which he shared in his usual self-deprecating way about how he and Mark Williams and John Teasdale all had to learn yoga when they were developing MBCT. MBCT was an adaptation of Mindfulness-Based Stress Reduction. The developer of MBSR, Jon Kabat-Zinn, had the genius to include yoga—a bottom-up approach—as a core component. As far as Jon was concerned, if they were to adapt MBSR, yoga was non-negotiable. It was too long ago to remember exactly what Zindel said, but I smile as I remember the picture he painted of the three of them learning yoga (somehow John Cleese comes to mind).

I spoke with Zindel recently about his latest book, *Better in Every Sense: How the New Science of Sensation Can Help You Reclaim Your Life*. The title alone bespeaks a clear evolution from top-down to bottom-up. Zindel and coauthor Norman Farb make a compelling research-based case for shifting attention away from habitual thinking and toward sensory input as a way to improve overall well-being and quality of life.

The authors succeed in not only making the figure/ground shift from thinking to sensing, but they also demonstrate through decades of research that "I think, therefore I am" might be more accurately restated as "I think, therefore I misconstrue." What they call the "house of habit"—and neuroscience calls the default mode network (DMN)—is designed to seek patterns, adapt to them, and make the world more predictable. This sought-after predictability comes at a cost: the premature foreclosure of novel sensory input and the failure to see reality as it actually is, which often locks us into unskillful habit patterns and limited ways of seeing ourselves and the world. To counteract this tendency, Zindel and Norman recommend sense foraging, which at its most basic level means diving more robustly and frequently into our sensory experience.

In our conversation, Zindel confirmed that sense foraging counts as a bottom-up practice, since it "reminds people about the primacy of sensation and offers a way of returning to the present moment. Maybe the grounding offered through sensory experiences can be a foundation for the mental attitudes of non-reactivity and even-handedness that are building blocks of equanimity."

There seem to be two key mechanisms in the therapeutic benefits of sense foraging: distress tolerance and decentering. These map perfectly onto equanimity, and they also provide the ground for receptivity to the novel to emerge, which in turn deactivates the DMN and takes us out of the "house of habit." Disrupting the DMN not only results in therapeutic benefits, but it's also at the heart of creativity and ingenuity. Zindel and Norman speculated that sense foraging gave us Newton's theory of gravitation (watching the apple fall) and Archimedes's buoyancy principle (stepping into the bathtub). I recently saw the play *The Lehman Trilogy*. In light of what I learned from Zindel, I was amused to notice that every Lehman brother's eureka moment (and there were many) was proceeded by the same refrain: "A gentle breeze wafted over him and then..." It was as if the playwright was suggesting that brainstorms like brokering commodities and investment banking were precipitated by moments of sense foraging.

Zindel and Norman were kind enough to share the following practice to give you a *taste* of sense foraging.

PRACTICE

You don't have to look too far to find ways of disrupting the default mode network.

For example, when you find yourself sitting down to eat, try eating with your non-dominant hand.

This may turn out to be one of the weirdest meals you ever eat, but weird can be another way of saying that you are stepping out of habit and expectation.

It could even be that this meal might feel satisfying in a surprising kind of way—noticing the effort it takes to explore, adjust, and accomplish the basic act of self-care that feeding ourselves represents.

What does it feel like to resist the self-criticism that comes from being slow and clumsy?

Maybe saying yes to the awkwardness of food dropping back on your plate, of taking longer than usual?

What about the flavors of the food when you eat this way, committed to sensing, exploring, and savoring each moment?

And when you come to the end of the meal, knowing full well that you can return to eating in your typical manner.

Is there a tinge of loss, of missing this quiet fumbling at mealtime, a temporary escape from habit, and getting to know a new home away from home?

Kaiut Yoga

For most of my adult life I've been a gym rat. Group exercise classes were my jam. I loved the music, the energy, and the accountability of keeping up with the group, and I've made lots of good friends at the gym. Over the years I graduated from high- to low-impact aerobics, from step classes to dance classes, and so on. When the pandemic hit, it forced me to stop taking classes altogether.

Of all the losses from the pandemic, losing my beloved Cuban salsa aerobics class probably had the biggest impact on my daily life. It was a source of joy, connection, exercise, stress management, and a big mood elevator. But desperate times called for desperate measures, so I put my trust in one of my dearest friends, Wendy Zerin, and began taking something called Kaiut yoga online from her. I had experimented with it a little before, but it didn't click at the time. I also had a love/hate relationship with other forms of yoga I'd dabbled with over many decades.

Although Wendy, a retired physician and longtime meditation

practitioner, was quick to let me know she doesn't consider herself a spokesperson for Kaiut yoga, she has been a dedicated Kaiut teacher and practitioner now for more than ten years. Francisco Kaiut is a Brazilian chiropractor who was motivated to explore many somatic modalities due to chronic pain from a childhood accident. This healing journey eventually led him to develop a new form of yoga, which he believes to be more closely tied to the original spirit of yoga. Yoga is now taught in gyms and rec centers, often with minimal mental training and a strong emphasis on poses that were created several thousand years ago. Kaiut yoga has deemphasized the fitness approach to yoga as well as radically adapting the poses, or asanas, to meet the needs of twenty-first-century bodies, which rarely sit on the ground or walk barefoot. It is also adapted to twenty-first-century nervous systems, which are assaulted with levels of stimulation inconceivable 100 years—let alone 2,500 years—ago.

So how is a Kaiut yoga class different from an Iyengar or Ashtanga yoga class? For one thing, the teacher never gets in the pose. All the instruction is verbal, and students are encouraged to practice a lot of the time with their eyes closed. This means that we're constantly being discouraged from trying to put our bodies into shapes that don't serve our individual needs. Have you noticed how competitive yoga classes can be? By contrast, in Kaiut, both safety and trust are inculcated through the warmth and knowledge of the instructor, inviting the student to enter into and remain in the parasympathetic mode of "rest and digest" throughout the class. As Wendy shared,

> The student can actually entrust their executive functioning to the teacher so they can really drop to a deeper level within, dropping not only below the level of discursive thinking, but dropping below the level of personality itself.

At first, and for a very long time, students are invited to practice in such a way that creates no discomfort nor triggers any defensive mechanisms. Eventually, students are invited to explore their capacity to tolerate the uncomfortable sensations that may accompany growth, without activating the sympathetic nervous system. In addition, all the poses are taught with multiple options so that it fulfills its promise to be yoga for *every body*. Stretching of muscles and soft tissues is de-emphasized and priority is given to finding and creating space in the joints, with the understanding that joint mobility is critical to functionality.

It's the fall of 2020, and my world is crumbling. My nervous system is overwhelmed, and the trusty spaces I go to for stress reduction and sanity are now dreaded germ-spreading and dust-gathering endorphin emporiums, and even my sitting practice can't hold what's swirling in my mind and body. I need to be still and to move at the same time. It feels like an impossible predicament, yet it's solved in Kaiut yoga.

For the past five years I've been practicing five to seven days a week and not once have I told myself I *should* practice. My mind/body/heart found what it needed and just kept going. Without even realizing it at the time, I needed a bottom-up approach to find equanimity when my world was on fire.

Although there are many degrees of trauma, it's rare to find oneself at the ripe age of seventy-two without having undergone several kinds of trauma. As my traumatized system was able to find equanimity through this form of yoga practice it caused me to reflect on the importance of soothing and stabilizing the body and nervous system when trauma is activated and how critical bottom-up approaches can be for certain people at certain times.

I had also been teaching workshops on equanimity and I began to notice a beautiful dovetailing between the concepts I was teaching and the instructions in Kaiut. So much so that I eventually invited Wendy to teach several trainings with me, combining many of the practices in this section of the book with Kaiut yoga. It was a perfect fit.

A DOZEN THINGS...
I've learned about equanimity through the bottom-up practice of Kaiut yoga

1. The body is a part of nature and governed by the laws of nature.
2. Aligning with my body aligns me with nature.
3. Aligning with nature helps me to find balance, trust, and connection.
4. The dynamic equipoise of equanimity can be discovered directly through the body.
5. This somatic experience of equipoise can become a template for equanimity in the heart/mind.
6. The body has its own intelligence that is trustworthy, grounding, and teaches me about the nature of reality.
7. Focusing on the body can disrupt the default mode network.
8. When my sympathetic nervous system is in hyperarousal, the body is often a more effective pathway through which to downregulate into a parasympathetic state.
9. In order to come into balance as the ground for equanimity, I need practices that bypass or short circuit the thinking mind (where I spend an inordinate amount of time).
10. Right effort—a term borrowed from Buddhism's Eightfold Path that means neither too little nor too much—is fundamental to

finding equanimity and can be readily learned through the body and then extrapolated to mental and emotional experiences.
11. There is an "inner geography" to equanimity that has physical correlates. (i.e., "Oh, equanimity feels like this in the body.")
12. These physical correlates can become a map to both find my way back to equanimity and to help me recognize it when I am there.

Wendy was kind enough to share the following brief practice which gives you a small sampling of Kaiut yoga. Please bear in mind that Kaiut is intended to be practiced as an integrated sequence of poses in a sixty-minute class and guided by a qualified instructor.

PRACTICE

Lie down on your back on the floor. Support your head with a firm pillow or a yoga bolster. Have your hips comfortably distant from the wall and extend your legs up the wall. If there's any discomfort in your spine or your legs, slide your hips even farther away from the wall so that your spine rests easily on the floor, and your legs can be straight but relaxed with your heels resting on the wall.

Have your eyes closed. Rest your arms alongside you on the floor. Have your nose slightly lifted, keeping the neck and the upper chest open...

Rest here and allow your system to settle for a minute or two.

Extend your arms straight out to your sides on the floor with your palms facing up, your elbows straight and your palms and your fingers spreading open widely. Feel the opening in your shoulders and your upper chest. If there's a lot of sensation here, stay with your arms extended out to your sides; if there's not too much sensation, hold the outside edges of your pillow or bolster with your hands. Stay there, or, if doable without causing any discomfort in your shoulders, slide your hands under the pillow/bolster and move your hands in the direction of interlacing your fingers under the bolster. Find the position for your arms which is highly sustainable, which delivers a clear but mild stimulation to your shoulders. Your nervous system must know that you are safe every step of the way in order for your practice to be of benefit. For this reason, there must be no pain and no struggle to maintain the arm position, no conversation in your mind about whether it's OK for you to be in the position.

As you rest here, feel your spine touching the floor. Don't try to change or correct the way your spine is touching the floor and don't search for any sensation in particular. Simply draw your attention to the sensations arising from having the spine on the floor. Stay with those sensations for a few moments, and, from there, allow those sensations to guide you into awareness of other physical sensations you might be experiencing. Feel the obvious sensations, and, as the system settles a bit further, feel the more subtle sensations as well. Allow the awareness of sensation to draw you into a state of deeper presence—mind and body united at the level of sensation. Rest here for a few moments.

Keeping your eyes closed, and keeping your nose slightly lifted, slowly turn your face to the left. Don't try to control the rotation; just turn until you come to the place where you naturally stop. Rest here and feel the sensations arising from the body in this shape.

If your neck is very sensitive, stay here resting at the end of the readily available movement. Otherwise, in a very kind, non-aggressive way, slowly deepen the rotation to the left a couple of times. Keep the nose lifted, keep the chin free and clear of the upper chest. Don't search for so-called problem areas; don't try to source or fix or push through any areas of rigidity you might encounter. Just be with sensations as they arise with a relaxed, calm but focused attention.

Slowly bring your face back to the center. Stay very present for sensations as you come back. Take your time resting in the center, allowing sensations to diminish.

Keeping your eyes closed, and keeping your nose slightly lifted, slowly turn your face to the right until you come to the end of the readily available movement, wherever that might be. Don't try to match the angle of rotation you had on the first side; just turn until you come to the place where you naturally stop. Rest here and feel the sensations arising from the body in this shape with a relaxed, calm, but focused attention.

If your neck is very sensitive, stay here resting at the end of the readily available movement. Otherwise, in a very kind, non-aggressive way, slowly deepen the rotation to the right a couple of times. Keep the nose lifted, keep the chin free and clear of the upper chest.

Slowly bring your face back to the center. Stay very present for sensations as you come back. Take your time resting in the center, allowing sensations to diminish.

Slowly release your arms down to your sides. Bend your knees and press the wall with the soles of your feet, pushing yourself back enough so that you can roll to your side. Take a long pause resting on your side, allowing the sensations in your legs from having them extended up the wall to diminish. Feel those sensations as they diminish.

Slowly roll onto your back. Taking your head support with you, push back from the wall enough so that you can extend your legs fully on the floor without your feet touching the wall. Have your legs separated wider than your hips and let the feet relax and fall out to the sides. Find a position for your arms that keeps the upper chest open but is completely restful, whether extending your arms straight out to your sides on the floor with your palms facing up, holding the outside edges of your pillow/bolster with your hands, or sliding your hands underneath the bolster and moving in the direction of interlacing your fingers under the bolster.

Stay here for at least a few minutes, taking care not to rush to analyze your practice. Simply rest, and allow the consequences of your practice to go in deeply, to be integrated into your system while you rest.

The Pull and the Power of the Body

There are many other body-based practices that can certainly strengthen mindfulness and may cultivate equanimity, such as tai chi, chi gung, and aikido. You may have noticed that somatic interventions have even found their way onto doctors' prescription pads. Compelling evidence now has doctors prescribing everything from walks in nature to forest bathing. UC Davis recently launched "Nature Rx," through which doctors and mental health counselors are now able to prescribe time in nature for students as a part of their treatment plans.

In moments of intense stress or challenge, the body may be the *only* accessible pathway back into equanimity. As we learned in chapter 5, balance, well-being, and insight do not come from repressing strong feelings. We need to "feel all the feels," ride the waves of strong emotions, and allow their chaos to disrupt the "house of habit" so we don't shut down or narrow our focus and miss critical input that might lead to an unexpected solution, a resolution of a conflict, or a brilliant invention.

Chapter 12

UPLIFTING STORIES

*The shortest distance between
a human being and Truth is a story.*

—Anthony De Mello

In our earliest history, when human beings gathered around the fire at the end of the day, it's unlikely they lectured each other or gave sales pitches. They shared their experience and wisdom in many forms—poetry, song, music, movement—and what they shared often revolved around a narrative. In other words, they told stories.

In the same way, when we're at home at the end of the day with our partners or families, we share stories, not PowerPoint presentations. When we get together with friends for a meal or a party, we exchange stories. When we stream a TV program or go to the movies or the theater, it's stories we're after. Stories are ubiquitous and universal. They bind us. They're the connective tissue that holds our circles of friendship and kinship together. They can also convey wisdom, large and small. And some well-chosen stories can breed equanimity. In fact, it's one of the main ways equanimity has been spread.

As I listened to stories from faith leaders and scholars from a variety of traditions for this book, one thing that struck me was the similarity in the spirit conveyed by these equanimity stories. Each was compelling—and helpful—in its unique way but they were also of a piece with the human appreciation for the resilience and flexibility, humor and simplicity of equanimity. A theme emerges here that will come up again and again as we go: as we understand what binds us in common humanity, we must also be careful not to paper over the real differences—in our cultures, our values, and our "positionality" in the world.

You will see that many of the stories collected here relating to equanimity focus on the dangers of getting caught in praise or blame, while others illustrate the importance of perspective-taking and appreciating changeability and impermanence.

Moses Maimonides and Sufism

Tom Block, whom we met in chapter 2, shared the following equanimity story from Maimonides. It sets the bar pretty high, as do many Sufi and Jewish tales. They often represent extreme tests of equanimity in the face of the vicissitudes, or worldly winds, and suggest a level of equanimity that's closer to enlightenment than most of us are aiming for. However, they point to an aspirational possibility that might serve as a beacon to steady our course when the going gets rough.

Equanimity, as a station along the path of spiritual realization in Sufism, was so respected by Maimonides that in a letter to a fellow rabbi, he retold this story about a Sufi adept to illustrate its importance. As the story goes, a great sage and philosopher was travelling on a ship and took a spot where people relieved themselves. Someone urinated on him, and he simply laughed. When they asked him why, he said that it demonstrated that his soul was at the highest level, since he did not feel disgraced.

One of the great hallmarks of Sufism is how it teaches through humor, beautifully illustrated in the tales of Mullah Nasruddin, a wise-man/fool whose antics range from harebrained to half-witted. Here are a few that offer an extremely down-to-earth version of equanimity.

Nasruddin and the Donkey

Going to town with his son, Nasruddin walked while his son rode their donkey.

Someone saw them and scoffed, "Lazy boy! Why must your father walk?"

So the son got off, and Nasruddin got on.

Farther down the road, someone else saw them and said, "Cruel father, making your son walk!"

So they both rode the donkey.

"Poor donkey, carrying two riders!" said the next person they met.

So then they both got off.

"Idiots!" laughed the next person. "At least one of you should ride the donkey!"

"Take note, my son," Nasruddin said. "There's no pleasing everyone."

Nasruddin's House Catches Fire

Nasruddin's house happened to catch on fire while Nasruddin was in the coffeehouse. One of his neighbors came bursting in to tell him the bad news.

"Nasruddin!" he shouted. "Come quickly! Your house is on fire!"

Nasruddin jumped up and ran to see what had happened. It was true: His house really was on fire. Flames were shooting up into the air and the whole structure was about to collapse.

Nasruddin, however, just stood there smiling.

"What can you be smiling about?" his neighbor asked.

Nasruddin replied, "Don't you see? I've finally gotten rid of those damn bedbugs at last!"

Nasruddin on Balance

Over time, Nasruddin became famous for his wisdom and learning. As a result, people came from near and far to ask him questions.

"I have a question, Nasruddin!" one visitor said. "Why is it that people choose to follow so many different paths in life instead of following the one true path?"

"It's actually for the good of the world that everyone follows their own path," Nasruddin replied. "Just imagine: If everyone followed the same path and ended up at the same destination, the world would lose its balance, tip over, and we would all plunge into the abyss."

On Hubris

Like Sufism, Judaism is endowed with a strong dose of levity. Jewish humor is often self-effacing and one of my favorite Jewish jokes offers a delightfully ironic take on pride.

A group of devout Rabbis are worshipping in the temple. One Rabbi stands up and says, "Oh G-d, I know I am worthless. I am nothing!"

After he has finished, a rich businessman stands up and says, beating himself on the chest, "Oh G-d, I am also worthless, obsessed with material wealth. I am nothing!"

After this spectacle, the custodian, who has been sweeping the temple stands up and also proclaims, "Oh G-d, I am nobody, I am nothing."

The rich businessman kicks the rabbi and whispers in his ear with scorn, "Look who thinks he's nobody!"

The Sleeping Man Under the Sea

There are countless Indigenous tales that illustrate the related concepts of humility, balance, and equanimity. As promised in chapter 2, the following is an abridged version of a Nanai story shared by Kiliii Yüyan. Unlike the previous stories, Nanai stories often involve complex situations in which there isn't a clear sense of right and wrong. The hero has to solve a problem and the solution isn't obvious. In this story, the protagonist shows us how creative solutions become possible from the non-reactive stance of equanimity.

> There is a giant who lives under the sea and is in charge of the supply of fish. At the beginning of the summer, he releases all the fish so they might feed the villagers. But he's fallen into a deep sleep. Brave Azmun has found the place where the giant is sleeping. He's crept up on the sleeping giant's chest. Will he wake him up and make him really angry? Release the fish himself? What to do?
>
> What he decides is to stand there and listen. The giant's snoring is very rhythmic and musical. So he pulls out a Nanai jaw harp, a shamanic tool, and starts playing. The rhythm wakes

up the giant, who is very pleased to be woken up by this beautiful, powerful music. He soon remembers that he has to let go of those fish.

That's the climax of Azmun's story, but it actually takes him many days to arrive there. He has a mission the village has entrusted him with, but along the way, there are many difficulties that could lead him astray. There are even sirens in the form of seals that try to lure him away and offer him a really good life. In fact, they are beautiful women wearing seals skins, and he forgets his mission for a while, but he has reminders, such as a knot that is tied on his sword, which helps him recall that his community sent him there with a purpose.

In a beautiful, mythical, and magical way, this story illustrates the power of equanimity in the midst of dire situations. Azmun doesn't lose all his composure in the presence of the fearsome sleeping giant, which then allows him to come up with a creative solution to the dilemma posed by waking up a giant.

Zen Stories

The Zen tradition is also full of stories that illustrate equanimity, a highly prized virtue in Japan and China. Here are a few favorites.

Su Dongpo

One day, Su Dongpo felt inspired and wrote the following poem:

> I bow my head to the heaven within heaven,
> Hairline rays illuminating the universe,

> The eight winds cannot move me,
> Sitting still upon the purple golden lotus.

Impressed by himself, Su Dongpo sent a servant to hand-carry this poem to Fo Yin. He was sure that his friend would be equally impressed. When Fo Yin read the poem, he immediately saw that it was both a tribute to the Buddha and a declaration of spiritual refinement. Smiling, the Zen Master wrote "fart" on the manuscript and had it returned to Su Dongpo.

Su Dongpo was expecting compliments and a seal of approval. When he saw "fart" written on the manuscript, he was shocked. He burst into anger: "How dare he insult me like this? Why that lousy old monk! He's got a lot of explaining to do!"

Full of indignation, he rushed out of his house and ordered a boat to ferry him to the other shore as quickly as possible. He wanted to find Fo Yin and demand an apology. However, Fo Yin's door was closed. On the door was a piece of paper, for Su Dongpo. The paper had the following two lines:

> The eight winds cannot move me,
> One fart blows me across the river.

The Prime Minister

The prime minister of the Tang Dynasty was a national hero for his success as both a statesman and military leader. But despite his fame, power, and wealth, he considered himself a humble and devout Buddhist.

He often visited his favorite Zen master to study under him, and they seemed to get along very well. The fact that he was prime minister apparently had no effect on their relationship, which seemed to be simply one of a revered master and respectful student.

One day, during his usual visit, the prime minister asked the master, "Your Reverence, what is egotism according to Buddhism?"

The master's face turned red, and in a very condescending and insulting tone of voice, he shot back, "What kind of stupid question is that!?"

This unexpected response so shocked the prime minister that he became sullen and angry. The Zen master then smiled and said, "*This*, Your Excellency, is egotism."

The Warrior and the Zen Master

A famous warrior had ridden through countless cities and conquered vast territories without ever having been defeated. Such was the horror he provoked in the people that when they learned that the army of the famous warrior was heading toward their country, everyone left—up to and including the rulers. Empty houses were left with pots still boiling on the stove, such was the flight.

Everyone fled but the Zen master, who lived modestly on the side of a steep mountain.

When the army took control of the capital, the famous warrior went to the cabin of the Zen master to see with his own eyes. Coming before him, he saw that it was a simple old man who'd not even stood up to beg for his life. The warrior burst out with insults.

"Old fool!" he said, drawing his sword, "don't you realize I am one who could cut you through without batting an eye?"

The teacher remained motionless and replied, "And I am one who could be cut through without batting an eye."

These are just a few stories from several traditions to illustrate the widespread respect for the virtue of equanimity. Enjoy them, return to them, collect a few of your own, make new ones up on the spot. Once you start looking for stories of equanimity, you may start to see them cropping up everywhere.

Chapter 13

BREAKING THE SPELL

*Life is a tragedy when seen in close-up,
but a comedy in long-shot.*

—Charlie Chaplin

Have you ever been completely caught up in a state of fear or anger—or even sadness—and you see or hear something that just cracks you up? And until that moment, you've no idea how thoroughly highjacked you've been by a narrative that's narrowing your perspective so much so that you can only perceive whatever affirms your story? It's as if humor breaks the spell you've been under, and suddenly you can see clearly again. Your body relaxes, your mind opens, and your heart softens. Ironically, the Latin origin of *humor* is a kind of fluid in the body—the opposite of rigid or contracted.

Swami Beyondananda (a.k.a. Steve Bhaerman) says that humor is a way of finding equanimity by restoring perspective. Having been a fan of the Swami's for at least thirty years, I was tickled to ask the master himself how mere humor mortals like myself might be able to access more humor-induced equanimity. Not only did he share wonderful stories from his own life, but he also provided some tools for finding the funny, using the ancient and time-honored tool of humor to restore balance, harmony, and sanity.

Most tools, whether contemplative or utilitarian, can be used with different motivations and will have different outcomes. This can be obvious, like a surgeon using a knife to heal versus someone wielding a knife with murderous intentions. It can also be subtle, like the same gift given openheartedly or given with an expectation of something in return. Same tools, different outcomes. Steve Bhaerman uses humor with the clear and conscious intention of fostering love and connection. And, as Joseph Goldstein always says, "Everything rests on the tip of intention."

This comes through in the personal and vulnerable stories he shares about how he has operationalized humor in his own life—because even the Swami can get caught in the spell of anger or impatience. As is true for equanimity in general, the key is not to never get caught; it is to recover more quickly. Here is the first of a few examples you might relate to from the Swami:

> I was involved in a very challenging negotiation, and things went from bad to worse. It's kind of like every word I said dug the hole deeper. And finally, I just slammed the phone down, and just then my wife, Trudy, walked in. She said, "What's the matter?"
>
> I said, "Boy, I wish I could do that conversation over again."
>
> And, I thought, *Why not?* I called the person back and said, "Hi, this is Steve Bhaerman. Did someone just call you pretending to be me?"

Steve calls this "self-facing" humor or "preemptive humiliation," and it is the very same quality of humility in the face of the worldly winds that all the spiritual traditions prize as a hallmark of equanimity. It is non-defensive without self-denigration. It's making fun without malice. UC Berkeley psychologist Dacher Keltner studied the benefits of teasing and distinguished teasing motivated by love or friendship from bullying. He has made an important case for "wholesome" teasing at a time when trigger warnings and political correctness may end up eclipsing levity.

It's clear that Steve and his wife, Trudy, enjoy the kind of playful teasing that Dacher's lab determined can shore up marital bonds. Here's one of many stories that Steve shared about how he negotiates conflicts with his wife:

> Trudy and I have an interspecies marriage. I'm a dog; she's a cat. Trudy is always the last one out of every gathering. I'm ready to go home. I'm the driver. I've got the leash in my mouth. This can get irritating. All of the things that you love about somebody can, after a number of years, devolve into "I hate that."
>
> If I leave someplace, even my own house, I have to make sure Trudy is in front of me, because if she's behind me, I'm there by myself. One day, we were going to the park to go hiking. I go to the car, and now I'm waiting. I'm sitting there and all of a sudden that irritation starts to surface. I'm getting angry at her for not being there, and I already know that nothing good is going to come from that.
>
> I've got to do something to change the energy. So, I go back inside and walk into the room, and there she is still getting ready. I say, "Trudy, I've been waiting my whole life for you, and you're certainly worth waiting for!" That broke the trance. She loved it. She's told the story many times, as a

marker of a great turning point in our mutual equanimity and in our relationship.

Here are some recommendations from the Swami about how to access greater equanimity through the doorway of humor.

The Cosmic Comic Magic Mantra

My wife and I were traveling in Europe and met a woman who worked as an interpreter. She told us the hardest thing to translate were jokes. They were often too language- and culture-bound. One day she was translating someone's talk, and he told a joke. She translated it, and the entire audience laughed. He was impressed.
"How did you translate my joke?" he asked.
"Oh, I just said, 'It's a joke—laugh,'" she replied.
Since then I have adopted that as a magic mantra whenever I have to face a "laugh-threatening" (note—not life-threatening) situation. I repeat "It's a joke, laugh" until I do. More often than not I will find something truly worth laughing about. Try this for a week. Then notice if you were able to get to equanimity—and maybe even find a surprising way to resolve the situation.

Practice Self-Facing Laughter

Have you noticed how many of our "problems" are really ego concerns—about looking good, being right, defending against criticism, and so on? One of the ways we create equanimity is recognizing these egoic triggers and literally laughing in their face. It's funny really how we try to hide certain things about ourselves, even though everyone knows them anyway! So . . . what are the things others criticize you about? What traits about yourself do you not like? Here are two ways to apply the magic of humor, and both involve exaggeration. Let's say you're shy. As a game you play in the comfort of your own mirror, make yourself ten times more shy. Do this until you laugh. Or, take the opposite approach. Be as bold as you can imagine, bigger than life and exaggerate that. If you're really daring, you will adopt one of these characters for Halloween or some other public event where no one knows you. Remember, self-facing laughter makes you defenseless. And no one can take offense at defenselessness.

Find the Joke Hidden in the Picture

I've also called this "pumping ironies," and others have called it the "reframe." That means applying the tools of the humorist—exaggeration, seeing things the opposite from the way they are, finding a surprising connection between two unrelated things—and using that to restore equanimity. I was feeling frustrated and a bit embarrassed about having two storage units. Then I came up with something that made me laugh and improved my mood right away. I said to a friend, "Ask me how wealthy I am?"

"OK, how wealthy are you?"

"I am *so wealthy*," I responded, "that my *stuff* lives in a gated community!"

Next time you're up against a frustrating or vexing situation—particularly if it's "chronic"—see if you can find a reframe to change your perspective.

Immerse Yourself in Cosmic Comic Consciousness

I read somewhere the average child laughs one hundred times a day; the average adult not so much. Laughter and humor are considered "neotenous" traits—those that make us more youthful (not to mention more useful). So consciously look for ways to immerse yourself in the humor that makes *you* laugh. When I moved to a new town some years ago, the doctor I met literally wrote me a prescription to go see a funny movie. He recommended the Irish film *Waking Ned Devine*, and I was glad he did. Be around kids and puppies. Apparently when we come in contact with friendly dogs, our body secretes these hormones called "puptides" that make us happier. You can also look at Jay Leno's funny headlines on YouTube (example: "Thieves Steal Burglar Alarm") or get a copy of his book. Oh, and do any of this with a friend and double your pleasure.

Commit Random Acts of Comedy

There's no humor more universal and more spontaneous than "situational comedy"—that which naturally arises in the midst of living life. If you've practiced the other four practices, you are more likely to be the one who sees the "funny" and shares it with others. If it feels right, you can find a book of short funny jokes you can insert into situations. For example, a friend and I were hiking and we encountered a woman who had a little dog. My friend said to her, "A guy goes to a psychiatrist and says, 'Doctor, I think I'm a dog.' The psychiatrist says, 'Well then get off my couch.'" The woman burst out laughing, and it shifted her mood.

Sometimes therapists and medical people use humorous postings as "people tenderizers" to help clients and patients relax. I remember one therapist had a cartoon blown up and laminated in her office. In the first panel, we see a boy being swept away by a raging river, yelling to his dog, "Lassie! Get help!" The second panel shows the dog on a psychiatrist's couch.

I've become increasingly forgetful in recent months. On a trip to New York City with my husband a few weeks ago, I managed to lose my hat, my phone, and a credit card in the course of a single day. Fortunately, I was able to retrieve all of them, but the temptation to both catastrophize and beat myself up was pretty powerful. (I am getting old, after all, and dementia could be looming.) Gratefully, I remembered a phrase the Swami used several times in our conversation. Whenever he shared a story about anybody messing up (including himself) he called them "silly geese." I remembered this and smiled and the spell was broken. Just like that.

Finally, as the Swami suggests, "When you're tempted to practice tantrum yoga, remember what we teach in the Absurdiveness Training Class: Don't get even, get odd."

Chapter 14

THE SERENITY PRAYER

One of the most profound and certainly the most economical equanimity practices ever written is the Serenity Prayer used in Alcoholics Anonymous.

It's easy to become inured to something profound that we've heard over and over again to the point it can sound like a cliche. The Serenity Prayer is no cliché. I include it here, as its own little chapter, just to give its wisdom pride of place. Feel free, of course, to begin the prayer with "God..." as it is typically taught, if you prefer.

Grant me the serenity to
accept the things
I cannot change,
The courage to
change
the things I can,
And the wisdom to
know the difference.

Chapter 15

TAKING REFUGE, FINDING EQUANIMITY

*There are two means of refuge
from the miseries of life: music and cats.*

—Albert Schweitzer

If we take refuge in something larger than our sense of self, our personality, our special likes and dislikes and preferences, it can help us find equanimity. Surrendering to something larger than ourselves, we realize that we don't run the whole show, and that can lead to peace—to a skillful response to universal conditions that deeply affect us all.

A refuge is a place where we're protected from the proverbial "slings and arrows of outrageous fortune," the ups and downs that affect and afflict us in the course of daily life. A refuge can be many things: an environment, a set of values, a religious affiliation, a divine figure, a community of like-minded people, or a beloved animal companion, to name a few.

This chapter caps off the practice section because the act of taking refuge typically involves a much bigger commitment than the thought experiments, reflections, and practices offered in the prior chapters in Part II. However, as with everything presented in this book, there are ways to dip your toes into the soothing waters of taking refuge that don't involve a commitment that overwhelms your already overwhelming life. Often, it's through the gradual accrual of moments of refuge that faith deepens and more serious commitments ensue.

It didn't occur to me to connect equanimity with taking refuge until a dharma friend shared a story that linked taking refuge with her ability to navigate a personal challenge with equanimity. With her permission, I share it with you as a beautiful example of how refuge helps us find ease and spaciousness, even in the middle of intense life circumstances:

> My big, beautiful, gentle giant dog, Koda, died. We took him to the animal ER, and, long story short, a mass had ruptured, and we had to euthanize him. He died peacefully in our arms.
>
> What followed felt remarkable to me. I experienced the full spectrum of emotions. I wailed when the vet first told us. I wept and chanted to Koda as he died. I miss him—but none of it felt like suffering. There was grief, but no fighting. I felt a kind of exquisite sadness I think many people can relate to.
>
> Despite its suddenness, I felt incredible gratitude that it happened as it did, and for all the kindness I experienced from the vet and my family and friends. I felt Koda's love and steadiness as if it had become a part of me, rather than him having been lost to me.

I went for a walk in the woods with our surviving dog and felt his delight at lowering his belly into the cool creek water on a hot day. It was so joyful and beautiful that I felt like I could take in the whole world in one bite. Today I feel heavy and sad, but not overwhelmingly, just naturally, so. It's been remarkable.

For me, this is 100 percent equanimity: feeling all the feels. And I know I have my practice to thank, brick by brick developing mindfulness and equanimity until it has become a part of me.

Three Refuges

Buddhism specifically speaks of three refuges: the Buddha, dharma, and sangha.

The Buddha is not only an enlightened being who lived 2,500 years ago, but he is also symbolic of the inherent potential for *any* human being to become enlightened.

The dharma is the truth of the way things are, beyond our projections and illusions.

Sangha is the community of others who are also on a path along with us, providing mutual support.

These refuges aren't only available to Buddhists. Under different names and guises, refuge forms a part of all spiritual and many philosophical traditions. This underlying unity and universal quality are unsurprising, since we all face the same existential dilemma: the worldly winds, the truth of impermanence, and the unavoidable nature of suffering.

Taking Refuge in the Buddha

In Compassion Cultivation Training (CCT), I teach a meditation where people bring to mind the most compassionate being in their lives. The idea is really to take refuge in what that person represents. A participant in my class once said, "I can't find a person, because people are just really problematic for me. What I've chosen as my compassionate image is this big, old comfortable couch that we've taken everywhere—even when we've moved to different countries. I refuse to get rid of it. I feel so safe and cozy on this couch." It reminded me of a TV show my daughter used to watch when she was little called *The Big Comfy Couch*.

I love that story. To me, it means we can take refuge in anything we find comforting, from the big comfy couch to our pets to a tree in our yard to the buds of springtime, and so on.

The most striking similarity among the faith leaders I interviewed was the depth of their connection with the divine, in whatever manifestation this took for them personally. In attempting to draw similarities between taking refuge in the Buddha and taking refuge in God or nature, it is helpful to go back to Bruce Alderman's perspective on homeomorphic equivalence from chapter 2. In each case, taking refuge leads to profound equanimity but the form it takes, including the practices, beliefs, and mindsets vary widely.

The primary difference is between theistic, particularly monotheistic traditions, such as Christianity, Judaism, and Islam, and non-theistic traditions such as Buddhism and Stoicism. While all traditions speak of the higher truth of the great mystery of life, theistic religions speak of this in the form of a god or gods, while non-theistic ones do not.

In Buddhism the Buddha was and is not a god but a realized or enlightened human being. This faith isn't necessarily connected to a deity or god, but to the realization of our own inner nature or Buddha nature.

For me, taking refuge in the historical figure of the Buddha has always been challenging. He feels remote and foreign. I wonder how many stories about him are true and how many are apocryphal. He isn't warm and fuzzy. He's more of a cold intellectual concept than a living being I can relate to.

Then again, I've always been allergic to the idea of God or any other divinity. Talking to each of the spiritual leaders rekindled a deep longing, envy even, for the ability to take refuge so fully and heartfully in divine presence. In fact, the most profound and moving aspect of those conversations was the personal sense of complete and utter trust and envelopment in divine presence.

Habīb described taking refuge in Allah as a womb of love, unconditional and exquisite. Rabbi Tirzah Firestone also described her relationship with G-d as unconditional love, profound, personal, and intimate. The closest I can get to this is taking refuge in the divinity of nature, in its beauty, intelligence, impartiality, capacity for renewal, lawfulness, complexity, elegance, and majesty.

Indeed, a common thread among the Indigenous leaders I spoke with was the sense of divinity immanent throughout the natural world. God was not confined to one figure but expressed through every creature, rock, stream, and star. Kiliii Yüyan has devoted much of his career to sharing the beauty and telling the stories of Indigenous people around the world, particularly in remote polar regions. He shared the following:

> For all our cultural distinctiveness, there is a thread that binds Indigenous communities, because we all have evolved in con-

cert with the land. Our connection with the land, with family, and with tradition are what make us Indigenous.

Michael Yellow Bird, a citizen of the Three Affiliated Tribes (Mandan, Hidatsa, and Arikara), shared touching stories of his ancestors' profound spiritual connection with the natural world as a place of refuge:

My grandfather, who passed into the spirit world some time ago, would go and sit, pray, and meditate, out in the Badlands in North Dakota when he was engaging in sacred ceremonies. During these times, he would reach a state where he was able to communicate with insects, animals, the lands, and the different spirits of the place he was visiting. Whether people believe he was able to do this is immaterial to me. What I know is that he was having a deep spiritual experience.

Taking refuge in a symbol of either awakening or divine love could take many different forms that range in degrees from repeating phrases to entering a religious order. For our purposes, here is a simple way to explore this refuge as a doorway into equanimity.

PRACTICE

Before you begin a challenging task or as you start a period of meditation or reflection, bring to mind an image and a felt sense of whatever symbol of love and wisdom has meaning for you. It could be God, Allah, Jesus, Buddha, Gaia, Quan Yin, your grandmother, or your eighth-grade English teacher.

Ask for their help, not so much to get a specific answer, but rather to release the sense of small self. Asking for help can be a way to access the perennial wisdom that gets occluded by the anxiety of the ego.

Often, it is sufficient just to call them up in your mind's eye in order to see through the delusion that you alone can solve the problems of the world without the help of all the sources of wisdom and support that lie beyond.

Taking Refuge in the Dharma

Dharma, the second of the three Buddhist refuges, carries many meanings. The most common one is "the Buddha's teachings," which are said to point to ultimate reality. Dharma, therefore, also means such reality. In other words, by taking refuge in the dharma, we take refuge in the truth of the way things are.

As opposed to the Buddha, the dharma is the easiest place for me to take refuge because, from a Buddhist perspective, this is simply taking refuge in the truth. Being inherently skeptical, I was never able to believe something just because it was written in a book, even or especially the Bible. Early in my Buddhist training, I was told not to believe in Buddhist teachings, but to look for myself. That changed everything for me. I was told where to look and then instructed to go see for myself what was true. Wow!

Insights experienced in this way are incontrovertible, even to the most hardened of skeptics.

Taking refuge in the dharma is taking refuge in any expression of truth and beauty. If we really broaden the frame, I take refuge in museums and the truth and beauty that I find there. Some people take refuge in the library and draw deep comfort from being surrounded by books. I have a friend who is a mathematician and mathematics is a refuge where he can find elegance, truth, and harmony. Refuge in the dharma can be understood as refuge in transcendent truth.

The Noble Truths and Marks of Existence

Buddhist teachings often begin with "the Four Noble Truths": the truth of suffering, the cause of suffering, the end of suffering, and the means to end suffering. When I first encountered these teachings, I was in my late twenties, lost, confused, and sad. I thought my suffering was a personal defect.

This shook me awake. You mean suffering is universal? Really? Before I heard this, I looked around and honestly thought everyone else was happy except me. Just hearing this was profoundly liberating. It was my first step in stopping the war with myself.

Suffering is also one of the three "marks of existence," another teaching that has been at the heart of my Buddhist practice. These marks of existence, which characterize the reality of our experience at all times, include impermanence and "not-self," as well as suffering. Impermanence refers to the inevitability of change and how, in light of this, all experience is inherently dissatisfactory (we can't hold on to anything). "Not-self" means that no unchanging, permanent self or essence can be found in any phenomenon. I particularly like Ruth King's way of framing these three characteristics of all phenomena: "not perfect, not permanent, not personal."

Michael Yellow Bird also talked about the role of suffering in his tradition; not only was the truth of suffering recognized but also its value in promoting equanimity:

> Suffering was seen as a virtue, and it was seen as one way to create within yourself a sense of humility, a sense of understanding, a sense of openness, a sense of compassion, so that you could experience someone else.

It's easy for me to take refuge in the truth of impermanence: it's simply so undeniably true! There is no part of my skeptical mind that can find any foothold in doubting this truth. Things change. Everything changes. What a relief to take refuge in this truth and not fight it.

For many years, I've been doing intensive silent retreats for periods ranging from ten days to three months. In this kind of practice, it's common to enter into altered states that could sometimes be rapturous and sometimes terrifying.

I developed the habit during a three-month retreat of beginning every sitting period by repeating silently to myself, *I surrender to the dharma*. This was a helpful mantra because it allowed experience to unfold and gave me the courage to stay put, whatever came up.

The dharma was trustworthy. It was and is the truth of unfolding life, and the tremendous effort of sitting these retreats was all about learning what that was: the great mystery of how life unfolds, whether we like it or not, whether we control it or not, whether it makes sense to us or not.

I couldn't surrender to a god because that just didn't make sense to me. I couldn't surrender to a person because people had proved untrustworthy. But I *could* surrender to the truth. The idea of not-self—the reality that our sense of "self" is really a fictional construct—can be more challenging. However, it's pretty obvious that when I'm suffering, I have a strong sense of self, and when I am happy and in the flow, my sense of self melts into the background and even dissolves. The more I take things personally, the more I suffer. There seems to be a direct correlation between less self and more happiness. Indeed, all traditions seem to agree that equanimity is inversely related to ego.

Chapter 6 offered a series of thought experiments to support you in taking refuge in these fundamental truths.

Taking Refuge in the Sangha

The third of the traditional three Buddhist refuges that help to develop equanimity is sangha. Sangha literally means a group or collection and, in this case, refers specifically to the group of people who are practicing the dharma along with you—like-minded people you can trust and rely on.

The simplest and most direct translation of sangha is "community." And there is no doubt that community can be a source of refuge. Judaism, Christianity, and Islam are particularly good at providing this through weekly prayer and children's services, numerous opportunities to connect on holidays, and charitable work as well as special interest groups.

When Buddhism came to the West from a variety of different parts of Asia, it was decontextualized, which made creating genuine community more challenging. The pursuit of meditation practice has often been more personal than communal. I've participated in various meditation and discussion groups over the years, but community needs an accessible and *ongoing* means of getting together. It needs the kind of binding, consistency, and commitment one can find in family, similar to the spirit of the Robert Frost quote: "Home is the place where, when you have to go there, they have to take you in."

Extended families used to provide this kind of commitment and connective tissue, but the nuclear family is increasingly isolated from this larger community. According to former US surgeon general Vivek Murthy, we suffer from an epidemic of loneliness in the United States, which has also been noted by health authorities across the globe. Community is vital to health. If we don't have access to a faith-based community, we may need to create one for ourselves. We may need a chosen family.

As I mentioned earlier, for many years I took Cuban salsa aerobics at my gym. That class became a community. I found my audiologist there. I got writing and parenting advice. We shared our joys and heartbreaks. When I was going through cancer treatment, the *salseras* brought me food. Book groups can become communities, as can organizations like the PTA or a gardening club.

Another way to find refuge in community is through the power of the mind. Though never a substitute for real-time connection, in meditation or prayer it is possible to bring to mind those who are sharing your experience in this moment, your friends or ancestors, or all those who are looking for truth, equanimity, and wisdom.

TRY THIS

Just for fun, close your eyes for a few minutes and imagine all the other people who are longing for a peaceful heart. Maybe they are reading this book right now, just like you! Maybe they are at the dinner table after a long stressful day, with unruly kids, afraid to turn on the TV and get more bad news. Maybe they are imagining you, as you sit here imagining them. Maybe, just like you, they are yearning for a way to meet the fullness of life with greater ease and balance. Imagine that your longing and your values connect you to people all over this planet, right in this moment, whatever their outward circumstances might be.

Two stories come to mind about the power and the function of community.

We had good friends who lost their five-year-old son to an accidental drowning about twenty-five years ago. The dad would pour his heart out to his circle of friends via emails. My husband and I were so grateful for these communications, which allowed us to share in the depths of his pain and incredulity. We felt privileged to bear witness to his raw, unfiltered feelings. It gave us a way to hold little pieces of the pain with him. Those of us who received these intimate windows into his grief became a little tribe, surrounding him virtually and metaphorically.

For thirty-five years I facilitated support groups for cancer patients and their loved ones. We were all licensed psychotherapists and met weekly for supervision. We didn't always talk about the participants in our groups, but occasionally the magnitude of the pain we witnessed was so great we needed the group to hold it with us. It helped tremendously. Big pain can't be held by just one person.

As Andrew Dreitcer (whom you met in chapter 2) put it, in

Christianity, "There's the notion that all the followers of Jesus are one body. Together, they form this body of peace, manifested in active, fearless, compassionate, courageous behavior to make peace. In that way, equanimity, the ultimate peace, is supposed to flow into the world and transform it in the image of God."

This community came to play a critical role in Andrew's life as his wife was dying. They brought food and offered support to him, his wife, and his two school-age daughters as he was caring for them while also managing a weekly airline commute for his work in another city. In fact, it was his deep sense of refuge in a presence greater than himself (which I liken to the Buddha principle), truth (dharma), and community (sangha) that allowed him to remain equanimous through an extremely tumultuous and challenging time. It was this equanimity that allowed him to conserve all of his energy for the Herculean tasks of caring for his dying wife and his two young daughters, and making his long commute to his job. As Shinzen Young discussed earlier, equanimity creates a kind of "superconductivity" in the system, such that no energy is wasted on the friction created through resisting, suppressing, or dramatizing the truth of our experience.

I've learned the hard way that the best is the enemy of the good when it comes to community. It's hard for most of us to create a perfect community of friends—spiritual, familial, or otherwise. Communities come and go. They solidify, they shift, and they dissolve. It's important not to grasp at ideals.

In the context of equanimity, community can function to strengthen our courage and sense of belonging, but it can also work against balance and clear seeing when a sense of tribal identity heightens polarization and out-group bias.

TRY THIS

If you are fortunate enough to already belong to a community organized around faith or shared interests, lean into this as a place of refuge. Consider getting more involved through volunteering or more frequent attendance at community events.

If not, experiment with stretching your definition of community. If you play bridge or mah-jongg or poker, play with the idea of including your partners as part of your community. If you often see the same people walking their dogs while you're walking yours, think of yourselves as a community of dog walker/lovers who live in the same neighborhood. The stories we tell ourselves matter.

And then see what happens next. Do you feel more inclined to initiate a conversation? Do you feel warmer inside when you think of them?

Just in the writing of this chapter I decided to reframe my weekly online mah-jongg game as part of my community, and I instantly felt more enriched and abundant.

Creating new communities from scratch can be off-putting because it feels like such a big lift. See what happens by starting small and extending your sense of community to groups of people you may be taking for granted or discounting.

As we've seen, the three refuges pointed to by the Buddhist teachings—Buddha, dharma, and sangha—take many different forms in a variety of spiritual and philosophical traditions. These are all places or states of mind wherein we can develop equanimity by taking refuge in something larger than ourselves. At the same time, it's critical to remember what Bruce Alderman shared in chapter 2 about homeomorphic equivalence and generative (en)closures: to taste the fruits of taking refuge in a particular tradition, some degree of fidelity to that lineage is required to realize the unique flavor of that gift.

Refuges in nature, meditation, or community enable us to see through and loosen the grip of an ego afraid of the buffeting of the worldly winds. From this place of refuge, we can access both the courage and the wisdom to stand up and make a difference.

Before We Move On...

As you've seen throughout the second part of this book, there are myriad ways to invoke equanimity. Whether you dip your toes into the cool waters of equanimity by reflecting on the phrases in chapter 9, immerse yourself a little more deeply with thought experiments or guided meditations, splash around with the Swami's tips for finding the funny, or plunge into the deep end by taking refuge, this timeless capacity can be accessed by anyone, regardless of your religion or belief system.

Likewise, the fruits of equanimity express themselves in myriad ways. In chapter 2 we explored the universal theme of nonreactivity to the worldly winds as an expression of equanimity. It can also invoke faith, peace, courage, and trust: all qualities that help us enact equanimity "in real life," as we'll explore in the next section.

Bear in mind that, although many of the practices from this section are quick and simple contemplations, they're radically subversive and your mind is likely to rebel. When confronted with reflections that challenge some of our deepest held beliefs about who we are and how best to function in the world, the mind can be expert at spinning narratives about the disasters that will befall us if we let go of certain worldviews.

To practice meeting the challenge of a shifting perspective, let's consider a few final thought experiments before we launch into bringing equanimity into the world, in the final section.

You might think, *If I don't take things personally, I will lose my edge. I might become a sap, on the one hand—and just let everyone get away with everything—or I might miss important feedback that could make me a better person.* Instead of succumbing to these fearful assumptions, how about trying it to see? Take a day, or a week maybe, and experiment with taking things less personally and see how it goes. What actually happens? Do you lose anything? Do you gain anything? What is really true?

You might worry that if you remember that all things come to an end, you won't be as motivated, or maybe you won't be able to fully enjoy sensual pleasures, or will fall into indifference. Check it out. Take a day or a few days and carry the truth of impermanence into moments of both pleasure and pain. Does the truth of impermanence diminish your pleasure? Does it increase your pain?

Naturally, the truth of suffering is pretty easy to turn away from. How could it possibly help to be reminded that suffering is unavoidable when there is already so much suffering in the world? Here again, experiment for a day or two and see for yourself. When you don't get what you want, do you ever rail against reality? Do you tend to blame yourself, or others or God or the universe? What happens if you see not getting what you want as a natural and universal part of existence?

In order to truly bring equanimity into a world on fire, we need to allow these deep-set, fear-based views that can govern our actions to loosen up.

PART III

Real-World Equanimity

Chapter 16

EQUANIMITY IN A WORLD ON FIRE

> My companion in its brightest month
> A diamond cool radiance
> Lingers above the horizon
> Reminding me (in the words of the poet)
> To care
> And not to care
> As all the earth-bound madness
> Engulfs our lives.
> Steady, faithful
> A light in the darkness
> As the day
> Morphs into night.
>
> —Joseph Goldstein, "Venus in the Western Sky"

Is the world really on fire?" Tim Ryan, former congressman and Senate and presidential candidate from Ohio, asked me this trenchant question. We'll hear from Tim shortly, but most of us (myself included) accept that assessment unquestioningly, given climate change and all the other components of the current polycrisis: global epidemics, multiple humanitarian crises and wars, increasing polarization, threats to democracy, and economic instability. The question for this chapter, for this book, and for each of us, is: What can we do about it?

How can equanimity help at this critical moment? What does equanimity look, sound, and feel like when it's expressed in the context of engagement and social change?

In the last chapter we explored the idea of taking refuge. The meditation teacher Ajahn Amaro interprets taking refuge in the Buddha as taking refuge in wise and loving awareness. After all, when someone asked the Buddha following his enlightenment what he was (Are you a god? Are you a wizard? Are you a man?), he replied simply, "I am awake." Amaro goes on to suggest that "when we are genuinely taking refuge in that knowing quality, embodying that Buddha refuge, our intentions and our efforts come into alignment with reality."

Aligning with reality is key not only in creating a sense of confidence and courage but also in ensuring the most efficacious actions in the world. Amaro calls this "unentangled participation." I would call it equanimous engagement. Gandhi called it *satyagraha*, nonviolent resistance resolutely grounded in the truth. You might also call it hands-on letting go. Whatever we call it, this quality of trust in the awakened heart has the potential to generate an unshakability, an inviolability, that provides genuine refuge from the storms of praise and criticism, gain and loss, comfort and discomfort, happiness and unhappiness.

Perhaps you, like me, feel a constant tug, a voice that tells you that you aren't doing nearly enough to fix this mess we're in. Interestingly, one of the words we use for this phenomenon, *scruples*, actually comes from the Latin *scrupulus*, which means a "small, sharp stone." Some say the philosopher Cicero first used the word analogously, to compare moral unease to a tiny, sharp stone in your shoe that is a constant source of irritation.

Writing this chapter has been like walking in shoes full of pebbles. What can I possibly say that will help you to help this world? Who am I to provide answers when I struggle almost daily with these questions myself?

I'm learning that the point doesn't seem to be to have all the answers all the time. No big surprise there. Yet, we can find inspiration that may help us to take one next step, however tentatively and uncertainly. I want to offer you examples of people whose stories or very way of being have inspired me, examples of how equanimity might manifest in the context of engagement with the problems we all face. There is no single formula, no one-size-fits-all. Not only do we vary wildly in terms of our resources and capacities, our individual circumstances and conditions change constantly, requiring us to change course over and over again. When I asked Roshi Joan Halifax, not only a Buddhist priest but also a longtime social activist, what her advice would be on how to maintain equanimity in a world on fire, she said:

> Equanimity is a balancing act. Like Philippe Petit on a high wire between the Twin Towers or Ruth Bader Ginsburg navigating opposing positions on the Supreme Court, or Tom Udall, our former senator from New Mexico, who said politics was a process of correcting course and finding balance in the midst of polarizing conditions. Sailing is an image I especially like. When you're facing the conditions of the sea and the wind and changing light, there's no way you can point your boat to the destination directly. You're constantly tacking, correcting course. That's what balance is: constantly making course corrections in response to conditions, many small and some big.

Although equanimity is a vital skill, it's not *the only thing* the world needs. The world needs our love, compassion, generosity, wisdom, energy, intelligence, creativity, and vitality to name just a few. Nonetheless, the more I learn about equanimity and reflect on the current state of the world, the more convinced I am that—of all these beautiful qualities this may be the medicine we need the most in this moment. In chapter 18, we'll explore the relationship between compassion and equanimity and how they mutually strengthen and enhance one another. In chapter 19, we'll look more deeply at the pebbles that remind us of our values and how integrity and equanimity are inextricably linked. For now, let's explore what "unentangled participation" might look like for a variety of people and from a variety of perspectives.

Hands-On Letting Go

One stunning exemplar of equanimous engagement is captured in the iconic photograph of Rosa Parks sitting in the "whites only" section of the bus in Montgomery, Alabama, in 1955. This small, unobtrusive, bespectacled forty-two-year-old woman dressed in a modest brown coat and hat, with her hands neatly folded over the purse on her lap, quietly changed the course of history.

She took her seat with silent determination and an almost uncanny sense of calm, in the midst of a chaotic, violent powder keg of social upheaval. She demonstrated the potency of equanimity, how it has nothing whatsoever to do with either physical strength or indifference. Rosa Parks sat calmly on the bus. She was neither reactive nor in denial. She cared deeply about equal rights, and she knew full well what she was facing. The very next morning, a citywide bus boycott was set into motion, which eventually resulted in the US Supreme Court decision, just one year later, declaring segregated buses unconstitutional. This was the power of profound equanimity.

In her book, aptly titled *Quiet Strength*, an inspiration for this book, Rosa Parks shares that her grandfather and her faith in God gave her the courage to sit quietly on that bus, in spite of the very real harm she knew she faced. Rosa Parks leaned deeply into the refuge she found in the church, her community, and her belief in human goodness.

The cultivation of equanimity helps us develop the ability to advocate effectively for the benefit of ourselves and others. Andrew Dreitcer found inspiration in the work of Reverend James Lawson, who taught many participants in the civil rights movement the principles and practice of civil disobedience—including how to be, in Dreitcer's paraphrasing of Lawson, "non-reactive [that is, equanimous] in the midst of people spitting on you and putting out cigarettes on your head."

Theologian Howard Thurman beautifully captured the role of equanimity in sustaining courageous engagement in his book *Deep Is the Hunger*:

> How may we work in the world courageously and intelligently, on behalf of a decent world, without despair and complete fatigue? What are the resources for personal rehabilitation and renewal? That we may be able to look out on life with its vicissitudes, the cruelty and transient joys, with quiet eyes and a tranquil spirit?

Those involved in the civil rights movement, Aizaiah Yong (Andrew's colleague whom you met in chapter 2) pointed out, were particularly impacted by "recognizing Jesus's own experience as an oppressed person." Aizaiah, too, was deeply impacted by Howard

Thurman's teachings about Christ's nature. Divinity was part of the community of the oppressed.

There are countless examples of equanimity leading to profound social impact. We could just as easily conjure up the image of Nelson Mandela walking out of twenty-seven years of imprisonment on Robben Island with seemingly unshakable equanimity and achieving something that had been inconceivable up to that point: the end of apartheid in South Africa and the emergence of an Indigenous-led government.

Paragons of virtue can be daunting, but they can also inspire. It's hard to imagine having that much courage and faith. Though I don't expect to embody that degree of equanimity (at least in this lifetime), it inspires me to know that human beings are capable of it.

Many such examples never make the news and, likewise, don't set the bar as high as these remarkable individuals. What they all demonstrate incontrovertibly, however, is that equanimity is not the same as passivity.

Mr. Paul

In a world on fire, every drop of equanimity counts and none of us can ever fully know the impact of our smallest deeds. In discussing this chapter with my wonderful editor, Barry Boyce, he recalled a small Jewish immigrant family stranded in a predominantly Protestant and somewhat xenophobic town in south-central Pennsylvania in the 1960s.

> In the little town I grew up in there was one small synagogue, only a handful of Jews. Among them were two brothers who

had immigrated from Hungary. They both ran dress shops on Main Street. My brothers and I, when we reached early teenagehood, worked for the older of the two, cleaning up and making boxes for customers' purchases. Mr. Paul was a quiet, unassuming bachelor, and as an old-world gentleman he addressed everyone with some form of honorific in front of their name. At thirteen, I was Mr. Barry. His life followed a simple routine. He and his brother had lunch and morning and afternoon coffee together every day. Mr. Paul conveyed amazing dignity; just being around him calmed you down. He'd obviously escaped some horrible things, which he didn't talk about. He didn't change the world, but in his small way he subtracted from its polarization and confusion and exemplified how to be a humble, kind human being. I've carried the inspiration of his type of equanimity with me throughout my entire life.

It's so easy to overlook, dismiss, or minimize the importance of not *adding* to confusion, hatred, and fear. And I doubt Mr. Paul could ever have imagined his presence would have such an impact on Barry that it would end up as a shining example of the power of equanimity in this book. In a case like Mr. Paul, the *absence* of something was even more important than the presence of something else. Circling back to Tim Lomas, the Harvard psychology researcher you met in chapter 5, and the shifting paradigms around what leads to human flourishing, it's also true that most Western societies have historically undervalued and overlooked qualities as subtle and amorphous as the absence of drama.

As I write this, many of us are finding that even the news channels we agree with are hyperdramatic and prey on our overtaxed nervous systems. As I cast about looking for reliable sources to keep me an informed citizen, I long for the quiet dignity of the Mr. Pauls of this world and pine nostalgically for the equanimity and gravitas of newsmen like Walter Cronkite or Edward R. Murrow (minus the patriarchy of course).

Soulfulness

About twenty-five years ago, at one of the first international mindfulness conferences, after giving a talk I was approached by a petite attractive woman who said that Jon Kabat-Zinn suggested we meet. She was tentative—although already a law professor at the University of San Francisco—and she was intrigued by the possibility of bringing mindfulness into the law. Fast-forward twenty-five years and this modest but dignified young woman has blossomed into a powerful force in the world. A recent keynote she delivered at a Mind & Life conference brought the audience to their feet, clapping, cheering, and stomping in awe and appreciation.

Speaking with Rhonda Magee about equanimity, I was struck by her thoughtfulness and the care that she brought to her speech, her appearance, and her listening. In my imagination, she has a lot in common with Rosa Parks: petite women, carefully dressed, dignified, committed to social justice, emanating quiet strength, and capable of rocking the world.

When Rhonda talks about her grandmother, her immense love and respect comes through with every word. Her grandmother is a signifier for a larger context—the environment she grew up in as a Black girl in the South and how she learned to find the strength to remain resilient in a world that didn't accept you as a full-fledged human being. While it may not have been called equanimity per se, what she was picking up on was deeply infused with it.

> My grandmother represented a broader set of Black social gospel teachings, which was also just part of the water I was swimming in as a young girl. It's about a way of being in relationship to constant struggle that I was born into in the South in 1967. It bears a lot in common with Martin Luther King Jr.'s and Howard Thurman's teachings, among others.

Equanimity for Rhonda is something that comes "out of or on top of an ethic of love, that we all already matter, out of the teachings and example of Jesus Christ." When your back is against the wall, when you're "existentially vulnerable," she asks, what can you rely on? She talks about a principle of responding creatively rather than reactively that she learned from her grandmother: "make a way out of no way."

In her own life, that credo has made a profound difference, as we see in her accounts of being othered in her book *The Inner Work of Racial Justice*. In our conversation, she referenced a story she once read about Benjamin Mays, one of the early presidents of Morehouse University—also one of Howard Thurman's teachers—who was reportedly slapped as a young man walking through the streets of Atlanta by a white judge, because, the judge said, he just looked like he thought he was too good. "That's radical vulnerability, existential vulnerability, when someone would slap you in the street for not being humble enough."

The equanimous response to even the severest of social injustice, Rhonda says, is to "do what we can and then let go and let be." For Rhonda, this "letting go" does not mean resignation to injustice. To the contrary, it means loosening our attachment to specific outcomes precisely so we are able to *continue* to sustain in the struggle for justice. Rather than being a hindrance to activism, equanimity is an essential skill that helps us "not get stuck in whatever it is we're drawn to work on." For example, she says, a Black person "cannot ignore the reality that your back is up against the wall most of the time, but you obviously can't live in that reduced space. And Black folk have figured out how, in the face of the effort to be reduced, you just are *not* going to be reduced."

And this not being reduced, in the Black aesthetic tradition, emerges as "soulfulness," which, Rhonda says, shares something in common with what we call equanimity. To define soulfulness, toward the end of our conversation, Rhonda turned to an icon of soul music:

> I'm paraphrasing, but when asked to define soul, James Brown, the Godfather of Soul, said that soul is the creative response to the kind of world in which you want to love everybody and feel love and feel like you belong and feel like you're making a world where love prevails, and yet everywhere you turn, you get the opposite of that.

You Have to Have Tools

I don't know Tim Ryan—we've met only twice—but he feels like a friend. I first met Tim at a reception in Berkeley almost ten years ago when he was a congressman from Ohio. After his inspiring talk at the Greater Good Science Center, my daughter and I were introduced to him. She was a college student, interested in doing an internship on "the Hill" in Washington, DC. He gave her his card and suggested she speak to his chief of staff. Six months later, she was working in Tim's office. He's just that kind of guy. He's direct, warm, unpretentious, and he gets stuff done.

When I started looking around for exemplars of equanimous engagement, Barry Boyce suggested I reach out to Tim. They actually *are* friends, and Barry had seen up close how Tim navigated the stormy seas of political life with grace, humor, and generosity of spirit. Tim's a big guy; he was a high school and college quarterback and he fills up the Zoom screen. And people who run for president can be both arrogant and intimidating. Tim is neither of these. From the first minute of our long leisurely meeting, he put me completely at ease, engaging each of my questions with a serious lightheartedness. He's comfortable in his own skin and he invites you to be, too. He's confident without arrogance, articulate without a trace of pontification, self-aware without preoccupation, easy-going without compromising clarity or principles.

Taking action out in the world inevitably brings attention, and with attention can come criticism, anger, and conflict. It challenges our equanimity in the face of the worldly winds of praise and blame, fame and infamy, among others. In contemplating the challenges that come with seeking change in a world on fire, I thought it would be valuable to consult someone who's put himself in the front lines for decades—and who has also sought to develop mindfulness, compassion, and equanimity.

Tim was elected to the House, representing his hometown region surrounding Youngstown, Ohio, in 2003. At twenty-nine, he was the youngest Democrat in the House. In 2012, he put out the book *A Mindful Nation*, where he talked about his own path to mindfulness and reported on programs to bring mindfulness into many sectors of society. After two decades in the House, he gave up his seat to run for the US Senate. Despite winning the Democratic primary with 70 percent of the vote, Tim lost in the general election to J. D. Vance. Tim started our conversation by talking about trying and failing in public.

Tim's approach to cultivating equanimity seems to reflect what we learned from the research discussed in chapter 4 concerning how equanimity doesn't mean *not having a response* but rather how readily we bounce back. Coaches taught him to understand that nothing is as good as it seems and nothing is as bad as it seems.

> When you apply that attitude time and again, you learn to not take failures personally, even when they're very public. So you can lose an election for the Senate, and get up off the mat and give a concession speech. I knew my kids would be watching. I couldn't go out there and be an angry asshole and then do equanimity a few days later. The bounce-back is also about letting go of ego, of pride, which is an anchor that drags you down.

Tim went on to share his own version of the Hakuin story from chapter 2. Remember Hakuin having a baby thrust upon him, only to be taken away a year later, and in each case his only response to the news was "Is that so?" As we look back on things that we thought were bad at the time, they can turn out to have had beneficial aspects in the end. "When I lost the Senate election," he said, "I was naturally very disappointed, but it has also meant a break from the current congressional food fight and an opportunity to really spend some time with my son, Brady, at an important stage in his life, something I wish I had had more of."

Tim's long view also includes always asking, *Is that true?*

> When you say "the world's on fire," I have to ask whether that's true. Is it on fire? Isn't it always? There are many bad things

happening, and many good. When you understand you can't be certain about what the outcome will be, you have less judgment, and with less judgment, you can maneuver the winds of change better. If everyone's stuck on the belief that things are really bad, that the world is on fire, the leaders and everyone else will have a jacked-up amygdala, and less capability to do what can be done, no matter how big the fire might be.

Above all, Tim emphasized the need for tools and disciplines, which he finds in contemplative prayer and other spiritual practices, and in the tools provided by coaches and therapists. At the same time, the tools don't magically get you what you want.

I'm a big believer in the Stoic philosophy—understanding what you can control and what you can't. You cannot control outcomes. That's ego. You can control your attitude toward outcomes, and for that you need all the tools you can get.

Feel all the Feels

I reached out to Kritee Kanko, whom we met in chapter 1 and again in chapter 8, because I was intrigued that a climate scientist believed that "feeling all the feels" was one of the most important things we could do to address the climate crisis. With a PhD in molecular genetics and microbiology, she served as a scientist at the Environmental Defense Fund for twelve years and has engaged in twenty-five years of cross-disciplinary research in several areas related to the climate crisis, food and water security, as well as biochemistry, microbiology, and geosciences. Kritee is also a Zen Buddhist priest who has shifted her professional focus to healing the polycrisis of our time through grief ceremonies, re-Indigenization, and ecodharma.

She began our conversation with a blessing:

> May your ancestors bless you.
> May your elders bless you.
> And may this book be of service.

Only three minutes in and I was crying. Kritee, too, teared up several times during our conversation. She is that rare breed of scientist who isn't afraid to cry.

In the hour I spent with her, not only did I learn about the power of nuanced engagement, but her presence and dedication also had a profound impact on me. For one thing, she helped me to understand important conceptual nuances like the difference between positionality (one's social position and power dynamic) and shared humanity. We may all be human and share basic drives and needs, but our access to resources of all kinds may be vastly different. Beyond that, though, it was her presence, her unflinching openhearted commitment to relieve suffering that left an indelible mark on my soul. It was impossible to look into the unblinking mirror of her fierce tenderness and embodied compassion and not ask myself, "Am I doing as much as I can with all the privilege I enjoy?" Without shaming me or preaching, Kritee invited me to look more deeply at myself, at this book, at equanimity, and particularly at what I hope to share in this chapter.

Just through her being, she challenged me to look deeply at whether or not I was hiding behind equanimity when more needed to be done. I've been grappling with that question. On the one hand, there is always more that needs to be done; on the other hand, even Kritee, with all her courage and commitment, has felt overwhelmed at times:

In the first three months after the Gaza-Israel conflict that started October 7, 2023, I got really depressed, Margaret. I have several Palestinian sangha members through my grief-circle work, and I just couldn't imagine what they were going through. I just forgot my own equanimity. It was so brutal seeing these images all the time, including of dying babies. I wasn't being useful with all my feelings of compassion and empathy. I would call it empathy without a sense of ground, of spaciousness. I was useless. If I had kept going, I would have become harmful to the collective field because either I was depressed or I was just so angry and just not able to make the changes. We need collective changes.

In her deep work on responding to the climate crisis, Kritee leads "grief and rage ceremonies," because she feels it's necessary for people to respond to the trauma in their lives in order to become available to do the work the planet is calling for. When we're subjected to trauma, she says:

> We are in fight, flight, and freeze. In that condition, it's hard to trust. When you don't trust, you can't have relationships, and we badly need our relationships to be a community, to have our power as community members, to ask our leaders for what we deserve, what other species deserve. That's why I say feel all the feels.
>
> I do the grief and rage ceremonies to really access the things that caused us trauma, so we can compost those traumas. Not that all traumas, all grief, can be completely composted; they stay with us lifelong. But a lot of them can be given air and water. A lot of garbage can be given air and water, so it becomes composted for our movement building.
>
> Feeling the feels is not about retraumatizing or dramatiz-

ing by remembering painful situations. You don't just keep digging up painful stuff without giving it some equanimity. Before and after each grief and rage ceremony, we do a lot of trust building, ground building, equanimity practices. Feeling the feels has to be done in a wise way, otherwise it's just more drama. That's good for theater, but it's not good for building up power.

Concentric Circles of Suffering

As I write, a brutal conflict continues to be waged between Israel and Hamas. The violence of the Hamas attack has been retaliated with even greater force. There is terrible suffering on both sides, heartbreak, violence, incomprehensible loss of life and dignity.

First, we watched aghast as young people and families were mown down and dehumanized in towns and kibbutzim just outside the border with Gaza. Since then, we've watched with horror the brutal decimation of the Palestinian people. The pain is so intense it is radiating out in all directions. There have been protests around the world.

What the people immediately affected by this crisis need is emotional, political, and material support—to simply survive the horror. As Weyam Ghadbian was quoted in an online dharma zine: "We don't want to hear you teach us about equanimity as they bomb hospitals."

It is never useful to *inflict* equanimity on anyone; to preach equanimity to people in the middle of war would be at best naive and at worst abusive. However, for those of us farther away, part of our job is to feel the suffering and find balance. It is difficult, very difficult. But if *we* don't, who will? Seemingly intractable age-old conflicts need every drop of equanimity this world can offer. Solutions don't become evident amidst reactivity and retaliation.

It is the job of every human being to both feel deeply the poignancy of crises such as these and not add to the drama. For those of us witnessing from afar, our job is to keep our hearts open to the suffering and bring as much clarity, balance, and wisdom as possible. Only from a place of wisdom can we then engage productively.

In the words of Roshi Joan Halifax:

> I believe that we have to let life into our lives, let others into our lives, let the world into our lives, let love into our lives, and also let the night into our lives, and not let the roof over our head, our knowing, our fear, keep out the moonlight. Altruism is exactly this permeability, this wall-less wilderness of the world, this broken roof that lets the moonlight flood our ruined house, our suffering world.

Though we're all affected, we are in concentric circles that radiate out from the center of a conflict. Those closest to ground zero are the ones who have the least access to equanimity. They're in survival mode. The farther out we get from the immediate circle, the more wherewithal we have to bring perspective to the situation. Because our nervous systems aren't overwhelmed, we have the responsibility to turn toward and face the suffering, feel it, and respond with wisdom, compassion, and the clarity that can only come from equanimity.

It is our job to bear witness, to feel deeply into the poignancy of the suffering, to understand that the whole world is involved in these internecine struggles. It is our job to have the freedom and spaciousness of heart to value all of life, no matter the tribe, to see possibilities beyond the binary of I am right and you are wrong. Those of us who are not in the immediate line of fire have the luxury to not get pulled into the drama and to remain clear-eyed. It may not seem to count for much, but being able to feel deeply while taking in the worst kind of information and not shut down makes us stronger, more creative, and available to find ways to be genuinely helpful when openings emerge.

And then, from this perspective of wisdom and compassion, we are each called upon to act—with whatever resources are at our disposal and in harmony with our deepest held values.

Safeguarding Joy

> Joy doesn't betray but sustains activism. And when you face a politics that aspires to make you fearful, alienated, and isolated, joy is a fine initial act of insurrection. Let us be fed by revolutionary joy.
>
> **—Rebecca Solnit**

A few years ago, I taught a residential meditation retreat in the north of Spain. One of the participants was a young woman from Chile, not more than twenty-five years old. We were practicing forgiveness, and after the meditation she shared tearfully that she could never forgive Augusto Pinochet because that would be a betrayal to the thousands of people who were tortured, executed, or "disappeared" under his regime.

As she spoke, I found myself imagining that I was one of those people, listening in as this lovely and vital young woman spoke passionately on my behalf, determined not to forget what happened to me and, tragically, conflating her personal happiness with loyalty to my suffering. I shared with her that, had I been a Pinochet victim, the last thing I would want would be for her to deprive herself of joy on my behalf. "Yes, I would love for you to fight for freedom and democracy in Chile. I would love for you to remember the horrors of dictatorship and to guard human rights and justice. But I see your life ahead of you. I would hate for your heart to be burdened by unforgiveness, for your joy to be dampened by your loyalty to my pain. This would not serve me." I'm not sure this perspective changed her mind that day, but there seemed to be an opening.

Kritee said to me that she sometimes thinks of equanimity as "grounded joy," not an out-of-control ecstatic joy but a joy that has gravitas. Whether you are on the front lines of a protest, volunteering in a war zone, or writing postcards from your living room, safeguarding joy is crucial for many reasons. Joy sustains motivation and energy by boosting morale and uplifting spirits. It fosters connection and encourages participation in social movements. Joy promotes open-mindedness and innovation, stimulating creativity, as well as empowering courage and risk-taking.

Civil rights activist Audre Lorde summed it up well when she said, "The sharing of joy, whether physical, emotional, psychic, or intellectual, forms a bridge between the sharers, which can be the basis for understanding much of what is not shared between them, and lessens the threat of their difference." I'm reminded of the rainbow pride flag, a joyful symbol that both unifies activists while also disarming detractors. No surprise that Swami Beyondananda chose a rainbow for his Zoom background. And interestingly he shared:

> Somebody once asked me when I was doing one of my humor workshops, what I thought the most suppressed emotion was,

and what came up for me was, oh, joy. That's the most suppressed emotion. It's always more acceptable to complain.

When joy takes the form of out and out humor in the face of suffering, it can feel sacrilegious, inviting how-dare-you looks. Yet it's often the lifeline that keeps us from drowning in a sea of despair. Not only does it break the spell of doom, it also reminds us not to take ourselves too seriously. Ram Dass and Paul Gorman edited a book of stories and reflections on service in 1985, and my two dog-eared copies have dozens of Post-it notes sticking out of the top. It's called *How Can I Help?* In many ways, it's been my bible, inspiring me, guiding me, and helping me clarify over and over again, the true spirit of serving others. A favorite story of mine from the book demonstrates the power of humor in the face of unthinkable suffering. It's from a clown who volunteered in hospital wards with terminally ill children:

> Some of us were setting up to show Godzilla in the kids' leukemia ward. I was making up kids as clowns. One kid was totally bald from chemotherapy, and when I finished doing his face another kid said, "Go on and do the rest of his head." The kid loved the idea. And when I was done his sister said, "Hey, we can show the movie on Billy's head." And he really loved the idea. So we set up Godzilla and ran it on Billy's head, and Billy was pleased as punch, and we were all mighty proud of Billy. It was quite a moment.
> Especially when the doctors arrived...

This same clown came up with the idea of bringing popcorn with him. When a child cried, he would dab up the tears with popcorn and pop it into his own mouth or into the child's mouth. "We sit around together," he said, "and eat the tears."

How Much Is Enough?

We all ask the question, in the end, "Am I doing all that I can?"

Around the first century CE, Mahayana Buddhism emerged, with a strong focus on compassion, embodied in the bodhisattva vow. As mentioned in chapter 8, this vow asks the practitioner to commit to relieving the suffering of all sentient beings before becoming fully liberated oneself. Now there's a lifetime of pebbles in your shoe! Taken together with this idea of scruples that Cicero bequeathed to us, perhaps the feeling of never doing enough isn't something we should expect to extinguish once and for all. Maybe we're meant to have that gnawing sense that there is always more to do in this floating world.

Accepting the mind's natural inclination to have thoughts can decrease rumination, and accepting life's inevitable unwanted experiences can create greater satisfaction. Just the same, accepting that we can never do enough may bring us some peace. If this is a given, and we don't try to get rid of every pebble, equanimity might reside in befriending the constant tug to do more. What if we could reframe our is-it-enough-ness as a wholesome form of restlessness, forever nudging us out of complacency?

Nothing can be more personal than the answer to the question "What can I do?" Tempting as it might be to judge others who seem to have more resources than we do, the answer to this question is both a moving target and the product of complex causes and conditions. A. J. Muste, a renowned pacifist and social activist, is best remembered for his unwavering commitment to peace and justice. One of the most iconic images of his activism occurred during the Vietnam War. In the early 1960s, Muste stood silently in front of the White House, holding a single candle. This simple yet powerful act of protest symbolized his steadfast opposition to war and his hope for a peaceful resolution.

A *New York Times* reporter approached Muste and questioned the effectiveness of his solitary vigil, asking, "Do you really think you are going to change the policies of this country by standing out here alone at night with a candle?"

Muste calmly replied, "Oh, I don't do this to change the country. I do this so the country won't change me."

Muste's quiet vigil became a powerful statement against the violence and a beacon of hope, inspiring many to join the peace movement. His actions underscored the impact of peaceful, nonviolent protest in the face of great adversity. We all have a different role here in the big mandala, based on our own circumstances; equanimity allows us to balance our need *to do* something with our need for peace, our need simply *to be*.

In *Conjectures of a Guilty Bystander*, Thomas Merton wrote:

> There is a pervasive form of contemporary violence to which the idealist most easily succumbs: activism and overwork. The rush and pressure of modern life are a form, perhaps the most common form, of its innate violence. To allow oneself to be carried away by a multitude of conflicting concerns, to surrender to too many demands, to commit oneself to too many projects, to want to help everyone in everything, is to succumb to violence. The frenzy of our activism neutralizes our work for peace. It destroys our own inner capacity for peace. It destroys the fruitfulness of our own work, because it kills the root of inner wisdom which makes work fruitful.

Chapter 17

PARADOX, POLARIZATION, AND...UNCERTAINTY

Some of us suffer from a debilitating mental disorder called irony deficiency. Seeing a doctor won't help, but seeing a paradox will.

—Swami Beyondananda

Let's start out with the bad news: Certainty is a delusion, and there is no ultimate safety. I'm sorry, this is a tough one. Among other profound spiritual paths, Buddhism has come up with some pretty great metaphors for facing up to our ubiquitous uncertainty. Trungpa Rinpoche, for example, is widely quoted comparing the path of awakening to jumping out of a plane and discovering you have no parachute:

> The bad news is you're falling through the air, nothing to hang on to, no parachute. The good news is there's no ground.

In this case, the parachute represents our ideas, beliefs, opinions, and mental constructs that offer a false sense of safety in a world that is constantly changing, uncertain, and full of contradictions. In fact, we go to great lengths to avoid existential insecurity.

Research from the Max Planck Institute in 2016 showed that subjects found it more stressful to have a 50 percent chance of receiving an electric shock than a 100 percent chance. It seems crazy, but I can relate to it, can't you? Dozens of adages attest to the universality of this conundrum including: Better the certainty of misery than the misery of uncertainty; a known evil is better than an unknown good; certainty of trouble is better than uncertainty of peace; the fear of the unknown is worse than the known danger; better the wolf at the door than the one lurking in the shadows, and so on.

Pema Chödrön's book *Comfortable with Uncertainty* has become a perennial seller across twenty-nine editions. She clearly hit a nerve.

Zen teacher and psychoanalyst Hubert Benoit used the vivid metaphor of an "icy couch" to describe the discomfort of confronting the raw truth of existence. One of his students, the great Zen teacher Charlotte Joko Beck, elaborated:

> Enlightenment is simply the willingness to live with what is, which includes the fact that we are often uncomfortable and uncertain. It's like sitting on an icy couch. If we are willing to just be with our discomfort, it eventually transforms us. This is where true peace is found—not in avoiding discomfort, but in being fully present with it.

While the existentialists I loved to read in college—like Sartre and Camus—implored us to get comfortable on the icy couch of reality, that bleak worldview was also the endpoint in existentialism. The good news, which Chödrön, Trungpa, and Beck remind us of, is that something quite wonderful lies *beyond* the icy terror of free fall: the discovery that there is no solid ground to crash into. But this can only be found by letting go of the need for certainty and learning to tolerate paradox.

In this case, the paradox is that by letting go of the need for security we actually become safer. Comfort with insecurity and uncertainty is the ultimate refuge. To embrace equanimity is to embrace vulnerability and sensitivity, and discover that the deepest peace comes from having nothing to protect.

Allowing Uncertainty

Certainty isn't an indication of truth.
—Joseph Goldstein

In 1927, Werner Heisenberg formulated the "uncertainty principle." Its formal definition is a little complicated at first glance, but its implications are profound. It asserts that we're inherently limited in the precision with which we can know certain pairs of physical properties—known as complementary variables—simultaneously. This principle, which became a cornerstone of quantum mechanics, revealed the limits of our ability to measure and predict the properties of particles at the quantum level. For example, if you try to measure the position of an electron and its momentum at the same time, you can't measure either one with certainty. The more accurately you measure one, the less accurately you measure the other.

While the uncertainty principle is specific to quantum mechanics, it provides a useful metaphor for the "unknowability" and unpredictability of the complex systems all around us. For some of us, having the scientific "seal of approval" on uncertainty as a physical law can help loosen our grip on the need to know and help us relax into the truth of impermanence and change. A related phenomenon in quantum mechanics also lets us know that the observer's position always influences what's being observed. What we see as "certain" is influenced by the perspective from which we're viewing it.

If we can't fully understand subatomic particles through highly advanced physics, how can we hope to discern the "right answer" to complex political quagmires? Who gets served when there's no room for a both/and position, and therefore no political discourse? When you can be either pro-Palestinian or pro-Israel but you can't be pro-Israel and pro-Palestinian? Who is served when anti-Zionism is necessarily anti-Semitism? There's no equanimity in the either-you-agree-with-me-or-you-don't approach.

I have a good friend who is a department chair at a prestigious university. In the months following the October 7, 2023, Hamas attack on Israeli citizens, she felt more and more compelled to take a stand as protests erupted on her campus. Like many Jews, she despaired of finding language that wouldn't amplify the conflicts, language that could authentically represent her own feelings, which were changing constantly in response to the changing circumstances. I saw this friend, this brilliant woman, this beautiful deep heart and soul, feel frozen at the impossibility of a situation that had become so polarized that any nuance was treated as intolerable.

When I spoke with her, I fantasized about what the world would be like if we gave our leaders permission to stand up and say:

"I'm not sure. I don't really know what the answer is yet ... I'm taking in more information."

or

"The situation is extremely painful. I don't want speak out of my perceptions, which are distorted by outrage and fear."

or

"I am sharing a position of uncertainty, and the only thing I have confidence in is that I am against unnecessary suffering."

Why do we insist that our leaders have to have the answer even if it is wildly wrong, premature, simplistic, or an underdeveloped view of the situation? Why do we force them into this corner? Why can't they be exemplars of a more balanced view? Of what value is false certainty? Can we make room for a leader to simply say "I don't know" and not fan the flames of polarization one way or the other? To show us that this, too, is a valid position and actually contributes to a more equanimous world. Naomi Rothman, professor of management at Lehigh University, has studied ambivalence extensively and has demonstrated that, not only is it an appropriate response to complexity, it can actually enhance leadership through promoting complex problem solving, cognitive flexibility, critical thinking, perspective-taking, open-mindedness, and intellectual curiosity.

Roshi Joan Halifax reminded me of her beloved teacher Bernie Glassman's three tenets, which were foundational to his form of engaged Buddhism:

1. Not knowing

2. Bearing witness

3. Taking action

It's so easy to skip the first step, yet not knowing provides a radical openness and a beginner's mind free of preference—a mind capable of taking in new information and much less vulnerable to confirmation bias. Not knowing requires *comfort with uncertainty*. And before jumping into action, there's still another step: bearing witness. This step is also easy to skip over as we try to avoid suffering. To bear witness *before* taking action is to be fully present to the immense joy and suffering of this world.

A beautiful example of a leader who embodied these tenets was Vinoba Bhave, one of Mahatma Gandhi's closest disciples. In 1951, Bhave began a pilgrimage that would take him more than fifty thousand kilometers on foot, through hundreds of small villages in India. After Gandhi was assassinated, a congress assembled to determine the path forward. Bhave was asked to lead the movement but, amidst the chaos and grief in the aftermath of Gandhi's death, Bhave did not believe he had the answers to lead. Instead, he decided to walk across India and listen to the needs of the people. He made it clear that he didn't have the answers and set out on his pilgrimage without a premeditated plan, other than bearing witness.

When he was visiting the village of Pochampally in Telangana, a group of landless villagers approached him asking for land. Their request moved him deeply, and he made an impromptu appeal to local landowners to donate some of their land. To his surprise, one of them immediately offered one hundred acres. This unexpected success inspired Bhave to continue his approach on a larger scale. The Bhoodan movement, which emerged organically from his interactions with villagers and landowners during this pilgrimage, resulted in the redistribution of millions of acres of land to the landless—all because he did not jump to certain answers and outcomes.

Exercise: Cultivating Curiosity

My life has been filled with terrible misfortunes, most of which have never happened.

—Widely attributed to Mark Twain (and Michel de Montaigne)

TRY THIS

The next time you find yourself worrying, get curious. What dire future scenarios are parading through your imagination? Without even a soupçon of judgment, ask yourself if you can really know what will happen.

If the answer is no, notice if the worry still has forward momentum, in spite of its inability to predict or control the future. For me, it often does.

Use the laboratory of your own heart and mind to empirically test whether a known painful outcome is instinctively preferable to the truth of not knowing.

Notice how worry serves the function of protecting us against uncertainty. Experiment with resting in not knowing.

PS: As a bonus, notice how much creative energy is freed up by doing this!

Opening to Paradox: Finding the Third Thing

Tolerating uncertainty is a very close cousin to tolerating paradox. They both require us to give up knowing something for certain. How can opposites both be true? Where is the solid ground to find a foothold? Can I surrender the need to reconcile opposing truths and find comfort in something larger and more mysterious than my ego can conceive of?

Yet another quantum physicist, Niels Bohr, pointed out that while a correct statement's opposite is a false statement, "The opposite of a profound truth may well be another profound truth. A great truth is a truth whose opposite is also a great truth."

Like equanimity itself, resolving paradox is a universal principle, emerging not only in quantum physics, but also in philosophy, psychology, and all major religions. Both Sufism and Kabbalah specifically incorporate paradox into their notions of the divinity. As the Sufi teacher Habīb Boerger, whom we first met in chapter 2, put it, there are two seemingly contradictory sets of names for the divine in Sufism: "a category of names that we might call beautiful or merciful, and names that we might call severe or majestic." These divine attributes are called *jamal*, meaning "beauty," among many other things, and *jalal*, meaning "majesty."

Habīb instructs the faithful to approach these different aspects of God through the gradual cultivation of equanimity:

> You see through God's revelation of giving life and God's revelation of taking life, God's revelation of strength, God's revelation of beauty, God's revelation of compassion, God's revelation of justice . . . and on and on. Obviously, you don't start out the path in that place of making no difference between the *jamal* and the *jalal*. That's something that happens through purification of the ego, through purification of the heart, and the heart's grounding, knowing, and being in proximity to God.

This is the same purification of ego or self that results from taking refuge in something that transcends our narrow, constricted sense of self.

In the Kabbalah, the tree of life symbolically represents how the universe came into being through the emanations of the Divine Oneness. As Rabbi Tirzah Firestone explained during our conversation:

> In the Kabbalistic tree of life, you have three pillars: the right hand, the left hand, and the middle pillar. [The right pillar] is expansive. It's lovingkindness, it's giving, it's boundless. The left is all about the limitations of time and space, the constraints of rigor. Both of those are completely necessary to create that middle pillar, which is the pillar of equanimity, of equilibrium, the synthesis, the middle way, the golden mean.

Christianity speaks of "the peace that passes all understanding." It's a familiar phrase, but what does it mean? What is it trying to point to? In one sense, it simply means that peace or equanimity cannot be understood through rational or logical means. Similarly, in Buddhist terms, peace or equanimity is a great and profound truth, or dharma, and therefore involves understanding and resolving paradox—because it deals with the vicissitudes, the contradictory and irrational worldly winds we encounter every day of our lives. How do you reconcile such opposites as praise and blame or gain and loss—without simply bouncing back and forth between them like a tennis ball? The fundamental paradox of equanimity, as introduced in the first chapter, is found in the contradictory words used to define it. How can you maintain the big picture (*upekkhā*) while standing right in the middle (*tatramajjhattatā*)? How can you find balance (*aequus*) while pulsing with aliveness and feeling all the feels (*animus*)? How do you find "the middle way"?

Rabbi Tirzah shed some more light on this for me. She's also a Jungian analyst, so she brought into our conversation Jung's idea of *transcendent function* when describing the necessity of tolerating paradox:

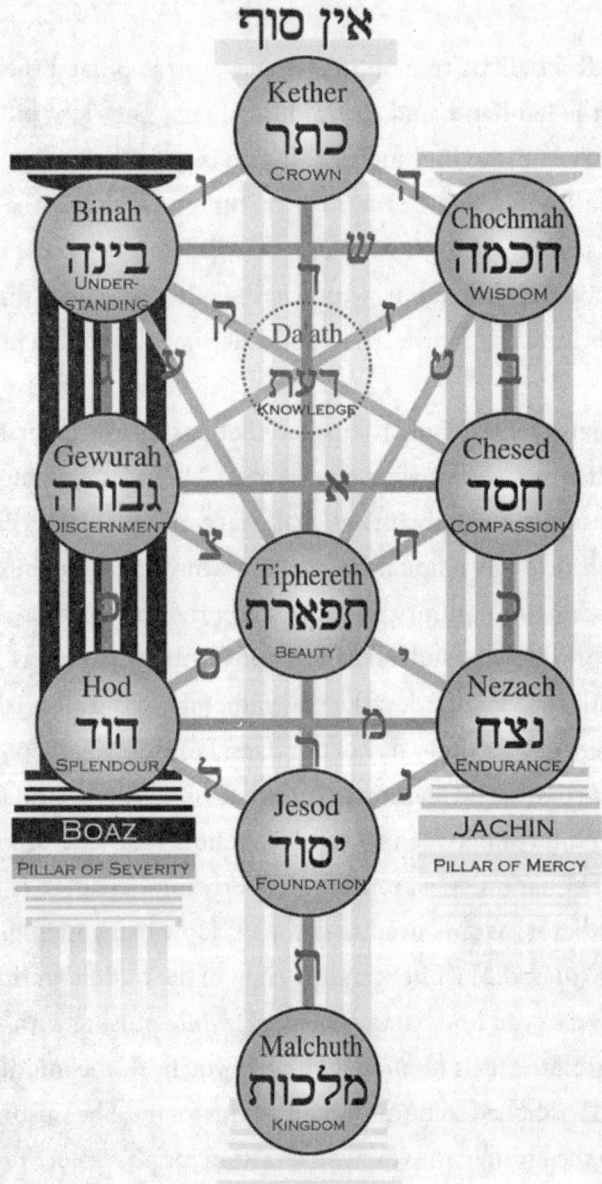

Jung talked a lot about being willing to hold tension with the opposites in your life. The opposites might be immense grief and the actual facts of what's going on. You just hold the tension, staying in those hard, uncomfortable places, and watch and wait for a transcendent function—a third thing, a new emergence, a new insight, a new dream symbol, a new something that happens. But we have to be willing to hold the tension.

This third thing, born from the dialectic of tolerating paradox, is equanimity. On a recent meditation retreat, Sylvia Boorstein described equanimity as synonymous with wisdom. Wisdom is born from the capacity to tolerate conflicting ideas/experiences and become bigger, rather than narrowing our options or becoming more hardened into binary perspectives. In a very real sense, equanimity is the synthesis of expansion and contraction, the two poles of the eight worldly winds we encounter every day.

Tolerating paradox is not simply relegated to the spiritual or esoteric realms. Ella Miron-Spektor from the INSEAD Business School and her colleagues have studied the value of the "paradox mindset" in promoting team creativity and balancing novelty and usefulness. Simply put, the paradox mindset helps us to switch from an either/or to a both/and framing of competing demands.

Enjoying the Dialectic

The word *dialectic* is closely related to *dialogue* and both come from words meaning to converse, to carry on a conversation. This kind of conversation is not about having a chat—which I thoroughly enjoy—

but rather about finding truth and wisdom by engaging with the polarities of life. It's celebrated in religion, as we've seen, and also in Taoism (think yin and yang), Socratic and Hegelian philosophy (where truth is sought through going back and forth), and now in psychotherapeutic mind-training disciplines, such as Dialectical Behavioral Therapy (DBT) and Acceptance and Commitment Therapy (ACT).

As you may recall from chapter 5, dialectic was a central tenet of Positive Psychology 2.0, in which Tim Lomas and others chose to *build on* and *be in dialogue with* positive psychology in order to transform the antithesis into a synthesis. In a book chapter on equanimity in psychiatry, Dave Vago, the contemplative neuroscientist we met back in chapter 4, mentioned the groundbreaking work of Marsha Linehan, who created DBT, which has been a great help to many people since the 1970s, including people I know. Mindfulness is a key component of DBT, helping to foster a more stable "lens" through which to engage one's thought process. And, as Dave Vago emphasized in the chapter, *radical acceptance* is one of the key skills DBT seeks to develop, and it strongly overlaps with equanimity. "Additionally," he writes, "the overarching dialectical approach, in which two seemingly conflicting ideas, feelings, or experiences can both be valid, can help to foster equanimity by promoting acceptance and non-judgment." *Vive la dialectique!*

ACT is another form of psychotherapy predicated on reconciling dialectic tension. Like DBT, it is part of the third wave of behavioral therapies, which combine mindfulness and behavioral therapy, providing a *third thing* beyond the two poles of acceptance without action and action without acceptance. In each case, paradox can only be resolved by making a leap into a different mode of understanding/seeing. Whether it's through religion, science, Socrates and Hegel, Jung, or psychotherapy, door #3 cannot be discovered by holding to the duality the fearful ego cleaves to in the face of uncertainty. Taking refuge in bigger perspectives and dialogues helps us to have the courage to leap into equanimity.

Loosening the Central Focus, Appreciating the Periphery

TRY THIS

Sit up as straight as you can in a chair. With your eyes closed, for a few minutes pay attention to your breath as it goes in and out. Now, when you're settled, open your eyes and look straight ahead with a relaxed gaze. After a while, without turning your head, place your attention on the periphery. Try to notice some details to the sides and up and down. Your attention will naturally come back to the center. When it does, go to the periphery again. Do that, allowing the back and forth to happen, for a few minutes.

You can notice several features of your perception in this exercise.

1. *Notice how difficult it is to maintain attention to the periphery and how attuned we are to a central focus.*
2. *Notice how we kind of deprioritize what's at the periphery, even though it may be quite important (it's where the escape routes are in a fire, for example).*
3. *Notice how it's possible to have both/and: a central focus and an appreciation of what is peripheral, the surrounding environment.*

Overcoming Polarization

Charles Lawrence is a former therapist and businessman who was inspired by his teacher, the famed mythologist Joseph Campbell, to become a medicine man. Part Blackfoot by origin, Charles was baptized by traditional Hopi elders, adopted by Lakota and Coast Salish (Musqueam band) elders, and acknowledged and accepted by Native American communities and Indigenous people near and far.

In a wild and freewheeling conversation, Charles shared mind-bending stories of his adventures, from accompanying a Hopi delegation to the United Nations to cocreating an ayahuasca center in the Peruvian jungle, decades before micro-dosing and ketamine clinics became ubiquitous. At eighty-nine years old, with sparkling blue eyes capped by unruly salt-and-pepper eyebrows, he exuded an enthusiasm I could barely keep up with. He had an insatiable appetite for in-depth dialogue and was my only interviewee to actually do "homework" before our call. Charles viewed our connection, arranged through our mutual friend Roger Walsh, as something more than coincidence. It was sacred, a perspective he seemed to bring to all his interactions.

When I asked Charles about how to find balance in polarized times, his answer quite surprised me. He spoke of his adopted mother, Carolyn Twangyawmama, a revered Hopi elder who spent her life working globally to promote the teachings and wisdom of the Hopi people, including prophecies about unity and peace for all people.

> I'm not one for peace. Don't talk to me about peace, not interested in peace, not interested in war. Grandmother Carolyn used to talk about this and she was a very direct little lady. She said, "If there was ever a government on the face of the earth that had it within its medicine, its powers, to have peace, we

would've had it a long time ago. If there's any religion that has that in its power, we'd have had it a long time ago." We've got to understand and admit that none of them have that gift, that power. If people want something different, they've got to come together and dream together, dialogue together.

He seemed to be suggesting we needed a *third thing*. It was shocking to hear someone say "Don't talk to me about peace," but my mind immediately went to the situation in Gaza and how the binary view of war and peace simply wasn't working, and never really had. The answer has to be a third thing. And Charles believes that this must be an *emergent* property, forged in the crucible of paradox and uncertainty, beyond what is known, and grounded in mutuality.

Relinquishing Attachment to a Fixed View

One way to understand the causes of escalating tribalism and polarization is to trace them back to their most primitive psychological roots: fear, uncertainty, and discomfort. When faced with existential threats, we regress to over-simplification, false certainty, groupthink, and confirmation bias—all of which are exacerbated through media and social amplification.

What were once considered fanatic views, so extreme they precluded conversation with someone on the other side, have now become mainstream. Destiny (a.k.a. Steven Bonnell) is a YouTuber and political streamer who is unafraid of debating extremists on both sides of the political spectrum. In a recent podcast interview with Sam Harris, he lamented the lack of bipartisan dialogue and momentarily woke me up from my own trance of delusional certainty:

Asking only one side to hold itself accountable is just really hard. We have a two-party system. We need that strong second party that can ground us out a bit. I remember debating conservatives and they changed my mind on some really powerful positions, six to seven years ago, and those people just don't exist anymore.

Hearing this, I remembered, again how deeply caught I am in believing that my political views are 100 percent right and opposing views are 100 percent wrong. I reflected on my cherished belief in the enrichment of diversity and wondered, with genuine interest, why this didn't extend to the political arena? Joseph Goldstein is full of pithy sayings that reinforce his understanding from his own deep practice and study. I've often heard him share this quote from the seventeenth-century Zen master Bankei: "Don't side with yourself." These four simple words have stopped me in my tracks more than once.

Many years ago, I engaged in a confrontation with my clinical supervisor in a cancer support program. She was an inspiring, powerful, and somewhat intimidating person, and there had been a disagreement between her and the clinical staff, of which I was a vocal member. I don't remember the particulars, except that we scheduled a "dialogue" during our supervision where the clinical staff would all observe and she and I would talk it out.

It was close to thirty years ago, but I can easily call up the anxiety I felt as the showdown approached. It felt like a therapist's version of an "appointment at dawn," only with an audience of my peers. Where was my second? It was as if my inner attorney was on a cocktail of cocaine and speed. As the hour of reckoning drew near, I spent more and more time preoccupied with making an irrefutable case for myself—

one that the judge and jury presiding in my mind would have no choice but to endorse.

And then I remembered what Joseph said and I made a U-turn. *Don't side with yourself.* In that moment, the rumination stopped. My intention dramatically shifted from winning an argument to listening to her perspective, engaging in a dialogue, and then seeing what happened next. There was still anxiety, but now it had a flavor of curiosity. This small movement toward greater intellectual openness and humility changed everything.

In an interview on Dan Harris's podcast *10% Happier*, Stanford psychologist Jamil Zaki shared research from his Social Neuroscience Laboratory on a phenomenon called false polarization. This is a form of cognitive bias in which we overestimate the extremity of the views held by members of opposing groups, heightening our sense of difference and in-group bias. There is no question that there's enormous polarization in the United States in this current period, he admits, but we "imagine it to be much more intense than it is, and we imagine the other to be much more extreme than they really are." For example, research shows that if you ask Democrats and Republicans "What does the average person you disagree with think?" the answer represents the most extreme 20 percent of the other side's views.

Zaki goes on to say that "we tend to imagine that way more people are extreme and violent than they really are." When we think this way, he says, "It forecloses on any possibility of productive conversation, of compromise, and of peace." In fact, in his lab they've found that more than 80 percent of Americans despise how divided we are. They wish there was more opportunity for compromise, which leads him to conclude that "there is so much that we have in common but we obscure that through our cynicism, which makes things so much worse."

MARGARET CULLEN

Dissolving Polarity Practice

TRY THIS

Should you happen to find yourself either in a trance of certainty about your "rightness" or in a fit of self-righteous indignation about someone else's "wrongness," ask yourself the following questions. (The last two are courtesy of Joseph Goldstein.)

1. *Could I be unwittingly engaging in false polarization?*
2. *Am I siding with myself?*
3. *Why do the other people feel the way they do?*
4. *What can I learn from this?*

This practice may seem like a drop in the bucket amid the swirling divisiveness we're surrounded by, but let's not forget people like the Mr. Pauls of the world who are not drawn into polarities even when pushed to the margins. He maintained his simple goodness and equanimous demeanor. His example had a big impact on my friend Barry at a young age. Hearing about it impacted me, and now the simple example of non-polarity may have impacted you. Not clinging to a fixed position can be contagious. On top of this, there's Zaki's compelling research that suggests our imaginations (or lack thereof) can quite literally compound the problem of polarization. Don't-know mind undercuts the pull to one side or the other. It lets you rest in paradox, and yes—in equanimity.

Chapter 18

BROKENHEARTED EQUANIMITY

> Overcome any bitterness that may have come because you were not up to the magnitude of pain entrusted to you. Like the mother of the world who carries the pain of the world in her heart, you are sharing in a certain measure of that cosmic pain, and are called upon to meet it in joy instead of self-pity.
>
> —Sufi master Pir Vilayat Khan

In much the same way that mindfulness and equanimity are inextricably linked and overlapping, so are compassion and equanimity. In fact, without equanimity, there is no true compassion and, without compassion, there is no true equanimity.

One simple way of understanding compassion is to view it as the outcome of love meeting suffering. In this context, love is the unconditional wish for happiness. Compassion adds a motivational element to love: the wish to relieve suffering. In each case, the heart is moved. Both the feelings of love and compassion touch, soften, and open the heart. In Habīb Boerger's words: "Everything is encompassed in the womb of love and compassion. It serves a purpose of returning to our true selves, of purifying the veils that lie over our innate primordial nature of goodness and oneness."

In chapter 1, equanimity was introduced as the last of the four immeasurables, following love, compassion, and sympathetic joy. Frank Ostaseski calls them the four flavors of love. Although they appear to be discrete categories with their own definitions and practices, they interpenetrate each other and share fundamental characteristics. One way of distinguishing them is through context. When love meets suffering, it becomes *karuṇā*, or compassion. When love meets the good fortune of another, it becomes *muditā*, or sympathetic joy. When love meets goodness, it becomes *mettā*, or lovingkindness. When love meets helplessness, it becomes *upekkha*, or equanimity. Each one of these qualities is boundless, indefatigable—you will never run out of it no matter how much you give it away. Each one is virtuous: When your mind is filled with any one of these qualities, it cannot be simultaneously unwholesome. It can't co-arise with greed, hatred, or delusion. Each quality embodies selflessness and consequently reduces self-centeredness.

Each has a near and far enemy. Here again, compassion and equanimity depend on one another. Without equanimity, compassion can fall into its near enemies of overwhelm and pity. Without compassion, equanimity can fall into its near enemy of indifference.

Compassion is a beautiful virtue. Giving, receiving, even witnessing compassion, elevates the human spirit and brings warmth to the heart and meaning to the universe. It was a privilege to study and teach compassion with Thupten Jinpa and colleagues for fourteen years, first through the Center for Compassion and Altruism Research and Education (CCARE) at Stanford and then through the Compassion Institute.

In the midst of immersing myself in compassion teachings, I began wondering why we didn't teach more about equanimity. What exactly is it and how does it relate to compassion? Why don't we talk about it more? Why doesn't it come up much in either the Buddhist retreat context or the secular contemplative world? Where are the books on equanimity? Equanimity seemed like the proverbial step-child of the four immeasurables—much less sexy than lovingkindness and compassion. And besides, underdogs have always been irresistible to me.

For the first few years of teaching workshops on equanimity, my students were primarily graduates of Compassion Cultivation Training, so we explored the relationship of compassion to equanimity in depth. What does equanimity have to offer compassion? What would be important for people who care about compassion to know about equanimity?

In a similar way to the spirit of the conversations I had in the first part of the book, the more I taught workshops on equanimity, the more excited about it I became (oxymoron notwithstanding). The most dramatic outcomes of these workshops were often in the context of family dynamics. The conceptual understanding of equanimity and the practice of loving without attachment helped parents to unhook from codependency with adult children and for adult siblings to repair emotional cutoffs. One man said he hadn't spoken to his brother in ten years and suddenly realized he was free to just pick up the phone and call him. He was able to see through an unexamined narrative that had limited him to being either cut off or a doormat and to see door #3 for the first time: He could love without expectation or resentment.

A Deeper Dive into Compassion

In order to explain how compassion and equanimity interrelate, it helps to get more fine-grained about what compassion actually is. Compassion is considered a virtue from the perspective of most religions, and defined by some psychologists as an emotion, but not by all. Although they don't necessarily occur in a linear progression, here are the micro processes that typically comprise a compassionate response:

1. An *awareness* of suffering

2. A *feeling* of being emotionally moved by suffering

3. A *desire* to see the relief of that suffering

4. A *willingness* to help relieve that suffering when possible

5. Often resulting in a *warm glow/sense of satisfaction*

What is this warm glow? And how is it that we can feel warm and fuzzy in the face of suffering? Research has shown that, even though compassion involves moving in close to suffering, the feeling of "How can I help?" is uplifting and energizing, even when there's little we can do to relieve the suffering. It feels good to connect with this fundamental instinct to care, to our basic goodness. According to Jamil Zaki, the Stanford professor of psychology we met in the last chapter, "humans are the champions of kindness." But why? Zaki's brain-imaging data shows that being kind to others registers in the brain more like eating chocolate than like fulfilling an obligation to do what's right (for example, eating leafy greens). Our brains have evolved to find it more valuable to do what's in the interest of the group rather than to do what's most profitable to the self.

This is in keeping with Dacher Keltner's revision of the classic Charles Darwin meme of "survival of the fittest." As a student of Darwin,

Keltner challenges the traditional view that human behavior is primarily driven by competition and self-interest, highlighting instead the importance of cooperation and altruism in human evolution. In his book *Born to Be Good: The Science of a Meaningful Life*, Keltner suggests that Darwin was misunderstood and a more accurate representation of his views regarding evolution might be "survival of the *kindest*."

In 2014, neuroscientists Tania Singer and Olga Klimecki, of the Max Planck Institute, published groundbreaking research that supported the idea that compassion actually feels good in the brain. Participants were instructed to respond either empathically (focusing on feeling the pain of others) or compassionately (focusing on lovingkindness and care) to videos of suffering people. Both groups were tested using fMRI imaging that records brain activity in real time. In keeping with Zaki's findings, the compassion condition stimulated the parts of the brain that are associated with reward processing and "warm glows" whereas the empathy condition lit up parts of the brain associated with pain and distress.

Going back to the granular definition of compassion, it might be fair to suggest that steps 1 and 2 are *empathic* responses (awareness of suffering and feeling moved by the pain) but, without the *motivational* components of steps 3 and 4 (motivation to relieve suffering and willingness to help when possible), we don't progress to step 5, the warm glow.

What Inhibits the Warm Glow?

Let's explore the factors that might keep us stuck in a purely empathic response to suffering. According to Penn State psychology researcher Daryl Cameron and others, there are several reasons we might get stuck in empathic distress and miss out on the warm glow of compassion.

In 2011, Daryl and his colleagues studied what psychologist Paul Slovic originally coined as "compassion collapse" and encountered results that seemed to defy intuitive sense. At the time, they used images of starving children in Darfur to provoke a compassionate response. It would be logical to assume that subjects would feel *more* compassion when shown images of eight or more starving children than they would feel when shown an image of a single starving child, but this isn't what happened. In trying to understand why we have more compassion for the image of one starving child than for many of them, this team of researchers have identified the following mechanisms:

1. Cognitive overload
   Shutting down: As the number of suffering individuals increases, people experience a kind of cognitive overload. If people *perceive* they will be overwhelmed and exhausted, they might take steps to disengage. It's easier to empathize with one person than with many and people may actively take steps to shut down empathy in order to avoid overwhelm.

2. Psychic numbing
   Diminished emotional response: Our emotional responses diminish when we are exposed to large-scale suffering. The impact of one person's pain can be profound, but as more people are added, the emotional response tends to flatten out. This can lead to a sense of helplessness or indifference, reducing the likelihood of compassionate action.

3. Perceived inefficacy
   Belief in limited impact: Another reason for compassion collapse is the perception that, when many people are suffering, one's ef-

forts to help will be ineffective. Daniel Västfjäll's research indicates that when people feel their actions will not make a significant difference (for example, donating to a large group versus an individual), they are less likely to feel motivated to act compassionately.

4. Moral distancing
   In-group–out-group dynamics: Daryl's work also touches on how people tend to feel more compassion for those they perceive as part of their in-group (for example, family, friends, or people similar to themselves) and less for those in an out-group. As the group of suffering individuals grows, it often includes more people perceived as distant or different, leading to reduced compassion.

5. Scope insensitivity
   Insensitivity to scale: People often struggle with scope insensitivity, meaning they don't proportionally scale their emotional response to the number of people suffering. Research suggests this lack of proportionality can lead to compassion collapse, where individuals feel a similar amount of compassion for one person as they do for many, which effectively dilutes the emotional response across larger numbers.

6. Self-regulation
   Managing emotional states: Daryl also explores how people sometimes intentionally regulate their emotions to avoid feeling overwhelmed or distressed. This self-regulation can lead to compassion collapse as people unconsciously or consciously dampen their compassionate responses to protect themselves from emotional exhaustion.

Tania Singer and Olga Klimecki's research suggested that compassion doesn't fatigue but empathy does. In light of this, they suggested that "empathic distress" might be a more accurate descriptor of what is commonly referred to as "compassion fatigue." From a Buddhist perspective, this designation makes a lot more sense. If compassion is immeasurable and unbounded, it can't get tired or run out, but it can get derailed.

In Daryl's research, compassion collapse happened for both naturally skilled emotional regulators, as well as with subjects who were primed to regulate their emotions. A second experiment in the same study showed that more highly skilled emotion regulators experienced greater compassion collapse. A third experiment showed the subjects who were explicitly instructed to down-regulate emotions showed compassion collapse, whereas those who were instructed to experience emotion did not. Down-regulating refers more to acting on the feeling of emotion on the inside rather than the outward expression of emotion.

Daryl summed it up nicely: "According to our theory, compassion collapse is not due to a limitation on how much compassion we can feel. Instead, it's the end result of people actively controlling their emotions."

How Does Equanimity Mitigate Compassion Collapse?

In his seminal paper on equanimity, which we discussed in chapter 4, our friend Dave Vago drew heavily on the research from Richie Davidson's lab on adept meditators. As you may recall, this research demonstrated that adept meditators are able to both *feel emotions more acutely* and *return to balance more readily*. They don't engage in "psychic numbing" or emotion regulation. It's not that they feel less; it's that they recover more quickly.

It's actually pretty confusing for many beginning meditators. What do you mean, *feel more*? I thought meditation could help me *feel less*. Feelings hurt. Meditation teachers and parents are often in the same position of affirming the bad news that life hurts sometimes and there are no shortcuts or magic bullets. Cultivating and practicing equanimity increases the capacity to feel, the capacity to tolerate the feelings, and the ability to regain equilibrium more readily.

In relationship to compassion, equanimity addresses each of the obstacles that Daryl's and others' research revealed:

1. **Cognitive overload:** The practice of extending love and equanimity to greater and greater groups of people is a targeted antidote. (See the "Just Like Me" practice in chapter 8, page 137.)

2. **Psychic numbing:** Indifference is the near enemy of equanimity. Understanding this key distinction is critical in avoiding the pitfall of psychic numbing. (See "What Equanimity Is Not" in chapter 1, page 20.)

3. **Perceived inefficacy:** This is where the Serenity Prayer (chapter 14) and the guided equanimity practice (see chapter 9, page 151) come to the rescue. They both remind us to accept the things we cannot control without forsaking our love.

4. **Moral distancing:** The "Just Like Me" practice referenced above also mitigates the tendency to offer love and care more readily to in-groups and show less concern for out-groups.

5. **Scope insensitivity:** The thought experiments (see chapter 6) are designed to remind us of the bigger picture. Equanimity allows us to stay connected with fundamental truths, including the scope of suffering, without withdrawing or turning away.

6. **Self-regulation:** This is subtle and can be confusing because emotion regulation has been touted as a good thing. However, Daryl's research indicates that it is often deployed as a strategy to *not feel* and avoid overwhelm. We also saw that Positive Psychology 2.0, Iris Mauss, Susan David, and other psychology researchers increasingly demonstrate the importance of honoring the full range of human emotions and the dangers of suppressing so-called negative emotions (chapter 5). Equanimity offers an alternative to the false binary of emotional suppression versus emotional overwhelm: feeling fully with a balanced and spacious heart.

In all of these ways, equanimity is what allows an empathic response to flower into compassion. In Buddhist teachings, as we've seen, the two near enemies of compassion are *overwhelm* and *pity*. Both of these derailments are averted with equanimity. It allows us to increase our ability to tolerate suffering so we don't collapse into the paralysis and self-preoccupation of overwhelm and to recognize our shared common humanity so we don't distance ourselves or assume a superior position through the aloof perspective of pity.

Brokenhearted Equanimity

It's the fall of 2020, and I'm in the depths of despair. Having suffered one debilitating episode of major depression twenty years before, I'm afraid of going there again—which only makes matters worse. I know what it feels like to consider suicide a viable option. At the time—two decades ago—the only thing stopping me from taking that escape route from my vortex of misery was my seven-year-old daughter. In my addled state, I determined that my husband could survive without me but my daughter could not. No matter how much I longed for that release, I just couldn't do it to her. I couldn't leave her with that legacy, that lifelong burden.

From the depths of that despair, I could never have predicted the places life would take me, the joy and fulfillment that lay ahead of me on every level, any more than I could have predicted a global pandemic that would collide with other factors to bring not only me but much of the world to its knees.

This is hard to write about. For one thing, thanks to the mind's brilliant capacity for selective amnesia when it comes to painful events, you may have forgotten the immense global suffering at the time of the pandemic. I'm sorry to remind you. It's also hard to write about because I'm sure many of you will have suffered far greater losses than what you're about to read. I have to remind myself of what I shared with the cancer patients in my groups: It's never helpful to compare our suffering with the suffering of others. Whether we determine it is greater or lesser, it separates us rather than connects us. Doing my best to put aside these concerns, I hope my story will be useful to you and will describe a possibility for an equanimity big enough to include the fullness of your life, excluding nothing.

From the ashes of my depression a remarkable and unexpected career blossomed that led me to opportunities I could never have dreamed of. I taught, traveled, met remarkable people, and was actually paid to do work I loved. My work had meaning and connected with my most deeply held values. It was an expression of my spiritual path. I made wonderful friends and was able to be of service to others.

At the start of 2020, I had already booked numerous international airline tickets to teach in exotic places that ranged from a retreat center in Brazil to a spa on the coast of Italy. The worldly winds had been blowing sweetly. I enjoyed encountering new cultures, having new experiences, and I relished the challenge of navigating international travel. It was exciting and enriching.

From March to May of 2020, this big, exciting world shrunk for me—as it did for everyone around the world—down to my Zoom screen, my bedroom, my kitchen, and my backyard. I thought I was sanguine about it, joking about every successive cancellation and acutely aware that mine were "first world problems." I marveled at my own ability to let go. That right there should have been my first clue. Through that spring and summer, most people around the world lived in fear of a deadly pandemic the likes of which we had never known.

By the fall of 2020, several factors conspired to mark the second lowest period in my life. September 20, 2020, was day 238 of the pandemic (remember, we used to count the days?) and fifteen thousand people had died in California. The world seemed to be falling apart. Here's what we were monitoring most days: Day # of COVID, total death toll, air quality, acres burned in California, countdown to November 3 (election day), businesses shut down, and county regulations for social distancing, to name just a few.

The sky wasn't falling, but it was truly surreal. Out my window in California the sky, if you could call it that, was a sickly red color—dark in midday, even though the sun appeared to be shining somewhere beyond the sooty haze. The climate crisis had contributed to the worst fire season in California's modern history with a record 4.4 million acres burned. We were in the midst of an extreme heat wave and rolling blackouts. If this were a dystopian Hollywood movie, it would look patently photoshopped.

We were on red alert to evacuate so frequently we just left a go bag by the front door for weeks. COVID was also raging again and we were all pretty scared. The presidential election loomed, and the prospect of four more years of rule by a person we regarded as a pathological narcissist with no respect for human decency, let alone the rule of law, deepened the sense of existential dread and despair. The holy trinity of the COVID deathwatch, raging fire storms, and looming fascism gave the world an apocalyptic quality that tipped the scales into massive overwhelm.

Right in the middle of this, the nonprofit I'd dedicated fourteen years to announced it was cutting all the programs I'd been involved with. COVID was taking its toll on for-profits and nonprofits alike. After the trips and plans had fallen like dominoes, this was my only remaining lifeline to the part of my world that had brought me so much meaning and purpose, not to mention income.

The worldly winds that had blown so sweetly just months before now brought gusts of deadly contagion, disappointment, and ravaging fires. The equanimity I was secretly proud of earlier in the year was wearing thin. Though I was still well aware that my circumstances were much more fortunate than most, inflicting perspective-taking on myself only resulted in suppressing the grief, fear, anger, despondency, and anxiety I was feeling.

When it seemed like things couldn't possibly get worse, they got worse. Something happened that pretty much undid me and stripped away any pretense I had of being able to use my years of practice and teaching to "transcend" my circumstances. I experienced a profound betrayal from a spiritual teacher I'd greatly admired. When that occurred, even the whole "spiritual enterprise," as it were, became tainted and suspect. Whatever I had learned in more than forty years of spiritual practice and deep inner work was neither strong enough nor sturdy enough to meet the challenge of the existential load I was facing in that moment.

I felt afraid and sometimes it escalated into terror. I felt sad; I cried a lot, and fell into despair for hours and days at a time. Meditation seemed like a hoax to me. The betrayal by the teacher was the proverbial straw that broke the camel's back, adding just enough weight to the load I was carrying to make me doubt just about everything I had learned. It's said in Buddhist teachings that doubt is one of the last hindrances on the path to enlightenment and the hardest to uproot. It's sneaky and pernicious. Its nature is to colonize the mind in such a way that everywhere you look there is uncertainty. Nothing seems reliable or dependable, including and especially ourselves.

Paradoxically, the only thing that is reliable in periods of doubt is doubt itself. When doubt is seen for what it is, its grip is loosened. The more we're seduced into believing in what is seen through the lens of doubt, the tighter its grip becomes. When it's understood as a lens that's skewing reality, we can connect with awareness of doubt itself— and awareness is never tainted by anything, and it is larger than everything.

Coincidentally, shortly after my professional and spiritual lives were torpedoed, I was scheduled to give a short talk and lead a practice on equanimity for the Mind & Life Institute (MLI). First mentioned in chapter 1, MLI was founded in 1987 to create dialogues between science and religion, particularly between the Dalai Lama and world-renowned scientists in an effort to integrate scientific research and theory with contemplative wisdom. I had attended almost all the public meetings as well as one private weeklong meeting at the residence of His Holiness in Dharamsala. MLI was the holy grail of everything I valued: scholarship at the highest level combined with the depth and integrity of contemplatives who embodied profound realization.

Though I had presented at a few conferences and MLI events, I always felt like an outsider and felt daunted by the contemplative and academic creds of this highly distinguished group. Talk about imposter syndrome. How could I give a talk on equanimity to *this* group when my life is in ruins and I can't even meditate? As a meditation teacher, my lifeline had always been authenticity. It's OK not to know, it's OK to have feelings, it's OK to make mistakes, as long as I can show up truthfully and openheartedly and in the service of the greater good.

Could I do that? What would it sound like? What would I say? Is it too raw, not yet ready for prime time? Should I just cancel?

Though I couldn't meditate, I had taken refuge in Kaiut yoga. Remember in chapter 11 when I wrote that I needed to sit still and move at the same time? I found refuge in the body and its simple animal*ness*. I could short-circuit the thinking mind and drop down into the world of *sen*sation, where things made *sense*, where doubt didn't have a foothold.

From this place of utter simplicity, not unlike T. S. Eliot's famous words at the end of *Little Gidding* . . .

> A condition of complete simplicity
> (Costing not less than everything)

I simply asked myself: *Is equanimity possible with a broken heart?*

And the answer was a resounding yes, from deep in my animal body. Yes, of course. Why wouldn't it be? And what use would equanimity be if it excluded this most poignant of human conditions?

The answer that seemed so elusive became immediately apparent. One of the phrases I'd worked with so many times over the years floated up from the depths of my being and refused to be contaminated by doubt: *This is how it is right now*. I didn't have to change anything about my circumstances in order to be equanimous or to talk about equanimity, even to the hallowed MLI audience. I just had to open wide enough to include brokenheartedness and existential despair in the field of equanimity. I had to lean into the truth of the moment and show up with my imperfect vulnerable self.

Vulnerability Is at the Heart of Equanimity

There are myths that inspire us and myths that tyrannize us. Under different circumstances, they might even be the same myth. There are several myths that plague me. They are both emblematic of the culture I was raised in, and they find purchase in the particular complexes or neuroses of my personality. In order to explain the primary myth that held me hostage as I contemplated the prospect of addressing MLI, we need to talk a little bit about my father.

My parents got divorced when I was five years old, and I didn't see my father again until my mid-twenties. He was a troubled guy by all accounts. Entered Cornell at age fifteen, tested off the charts on the army intelligence test, a very smart man. Our few mementos of him included a complete library of all of Freud's written work. The explanation our mother always gave for why he never saw us again was that he had decided, with outrageous hubris, that if Freud could psychoanalyze himself, so could he. She told us, once we were old enough to make sense of this, that his attempt to look at himself was so painful that it resulted in complete denial that he'd ever been a husband and a father. That he had "split us off" as if we had never existed in the first place.

I spent much of my childhood fantasizing about a fairy-tale reunion with my father, one in which he would be dazzled by the amazing daughter he had and we would live happily ever after. When I did find him after a lengthy search, instead of the fairy princess meeting her king, I saw a fairly broken man who had retreated into a small world of bitterness, paranoia, and self-preoccupation.

I was too young to accept his shortcomings and fully digest how paltry the reality was compared to the fantasy I had been carefully feeding and tending to all those years. So, instead, I invested a lot of energy in trying to get really smart men to respect me. There are several universal aspects of the myth that were deluding me. One was the *if only* conundrum: If only I could prove to a very smart and aloof male figure that *I* was smart, strong, and beautiful enough, all the pain from my father's abandonment would be vanquished. And, of course, in order to do this, I had to appear invulnerable. I couldn't show weakness, I had to be irresistibly perfect. In other words, I couldn't be myself.

To open the door to vulnerability was to let loose a floodgate of loss and insecurity that would show the world, once and for all, how deeply damaged I was. I would never find a substitute dad that way. Another pernicious and particularly American myth unconsciously shackling me was the idea that I could fix what was broken. And, perhaps more to the point, that brokenness was wrong, shameful, and demanded to be fixed.

If you're alive and human, your heart will break over and over. Your body will break, too. Bump your arm and you will get a bruise. Even with the miracle of modern medicine, long healed broken bones can sing a siren song in the right weather, reminding us of the fissures that remain beneath the surface, reminding us of our vulnerability.

MLI represented the fantasy ivory tower where my fantasy idealized self could meet many fantasy idealized versions of my father and all would be well for ever after. The gift of this particular moment was that I was too broken to hide my pain and brokenness. The yawning abyss between the fantasy version of myself—the one I hoped to present to the MLI community—and the truth of my actual pain and vulnerability was just too big.

Opening to the fear of presenting on equanimity to that audience from a place of brokenheartedness gave me the courage to write this book. I saw once again that the sky didn't fall. I reconnected with my spiritual practice and found that it could hold me after all. Equanimity allowed me to face my shameful need to reenact the tragic scenario with my father and hold it with compassion—a compassion that allowed me to look beyond the binary of victim/perpetrator with my spiritual teacher, who was in fact the last in a series of brilliant, powerful men whom I hoped might substitute for my father and help me paste a happy ending onto my sad story. It's pretty tricky these days to suggest that victims have any role whatsoever in their own abuse, and this can become yet another way we get straight-jacketed into binary views that poison the social discourse.

I felt, deep in my bones, in my spiritual DNA, the relaxation and strength that comes from opening to the truth. All of the metaphors I have shared in this book—the sensitive calibration of the gyroscope, the ballast of the keel, the grace of the surfer, the quiet wisdom of the grandmother—were suddenly available to me. Of course, they had been there all along, like Dorothy's ruby slippers and the Buddhist belief in basic goodness, much like the sun shining somewhere behind the ghoulish red haze of ash outside my window.

Chapter 19

CONNECTING THE DOTS: INTEGRITY AND EQUANIMITY

Come, come, whoever you are. Wanderer, worshiper, lover of leaving. It doesn't matter. Ours is not a caravan of despair. Come, even if you have broken your vows a thousand times. Come, yet again, come, come.

—Jalal al-Din Rumi

Way back in chapter 1, I promised we would learn more about the dysfunction that arose in my relationship with my mom. This is where we learn about that wild ride that brought me the kinds of ups and downs tailor-made for developing equanimity—or losing your shit altogether. Welcome aboard.

In my mid- to late twenties, the pain and unresolved problems from my childhood began to surface. I had very few healthy coping skills at the time and it often felt like an underground tsunami was threatening to break through the fragile surface of my life and persona. The eruption of emotion I described in chapter 1 was just one of many such episodes over the course of numerous silent retreats, years of individual therapy, and deep work with psychedelic medicines. I had a basketful of troubles, and it took many years and many different modalities to find my way to health.

My main coping strategies at the time were shoplifting and then numbing myself through overeating. The shoplifting had a number of "perverse benefits" or secondary gains: It was exciting (and, therefore, distracting); I felt a momentary high at "getting away with something"; I felt a momentary high at getting the *thing*, the object of my desire; I could fulfill a neurotic need to prove to myself just how unlovable I really was.

I also came by it honestly (with all the irony that carries). My mother was an unabashed shoplifter. She would brazenly slip anything she fancied into her capacious handbag. In the beginning, I remember hotel ashtrays and keys, the kinds of things you can more easily rationalize. Although I'll never forget checking out of a grand hotel in Lisbon and the manager chasing our taxi because my mother had taken one of those old-fashioned hotel keys that has a ten-pound brass fob attached to it. I was sixteen and thoroughly mortified.

In the convoluted inner workings of my mother's mind, this behavior was OK because:

- She was entitled to these things.
- They were "small" (even the eventual nighties and necklaces) and therefore inconsequential.
- She wasn't *hurting* anyone.
- They wouldn't even miss it!
- It was a "victimless" crime.

Although I always knew it was wrong, I had my own compulsion to reenact it . . . until I didn't. It was a dark secret that led me to countless hours of profound shame—and an eating disorder. The arousal from my petty crimes would invariably lead to cycles of binge eating. The process was compulsive, driven, and unconscious. I ate privately, often ice cream or junk food, to a point well beyond satiety, an addictive behavior driven by emotional avoidance. Though I was never bulimic, I would often swing to the opposite extremes of over-exercising or over-restricting my food intake in a desperate attempt to maintain my weight.

As I think back to that young woman, I can hardly believe it was me. But when I close my eyes, I can find her inside and my heart floods with compassion and tenderness for her pain, shame, confusion, and isolation. One of the most impactful and enduring things Marshall Rosenberg said to me, which guides me to this day, is that "all unskillful action can be understood as a tragic expression of an unmet need." This perspective makes it so easy to feel compassion for my younger self, trying desperately not to feel the pain and confusion swirling inside. And it helps me feel compassion for my mother, who yearned to be happy and loved and often behaved in ways that brought her the very opposite of what she longed for. I even feel a little awe that I have been able to climb my way out of that morass in a way my mother never could.

Connecting the Dots

The first and most important step I took was to attend a meeting of Overeaters Anonymous (OA). The lessons I learned there have stayed with me to this day, and that's why I included the simple and exquisite serenity prayer as its own chapter. It was my first equanimity practice. I left out the part about God, because that never worked for me, but the "moral inventory" was life-changing. As I said in introducing the prayer, please feel free to include God in the prayer if that works for you. If nothing else, I hope you will take from this book the fact that equanimity can be accessed *with* or *without* religion, *with* or *without* God.

There is so much genius within the twelve-step programs, but the part I would like to focus on here, and the reason for sharing my painful personal story, is the influence ethical behavior has on equanimity, and vice versa. Though it may seem obvious to you in reading my story, I had no idea my shoplifting was driving my overeating until I joined OA. Unskillful behavior—behavior that's misaligned with our personal values—creates agitation, rumination, dis-ease, imbalance, addiction. In short, *everything that undermines equanimity*. Through the combination of taking my shameful behaviors out of the "closet" and the moral inventory (step 4), I finally began to connect the dots. I admitted what I was doing to myself and others. I returned things I'd taken (now that took some guts—I will never forget the overwhelming urge to flee as I approached the young woman at the counter in a boutique to return the fancy belt I had stolen) and I stopped shoplifting.

A Moral Crisis

One personality trait I shared with my mother was rebelliousness. She rebelled against her Orthodox rabbi father and eloped with a communist intellectual, severing her ties not only with religion but also with her entire family. I, too, enjoyed breaking rules more than following them. On one level, our petty criminal behavior was an act of defiance, of getting away with something, of resisting authority.

For people like Mom and me, the moral edicts and commandments of religion were just another set of rules to rebel against. I needed to learn for myself, often the hard way. OA wasn't my only opportunity to determine if principles were morally "right" just because religion said so, or because I'd tested them in the hard knocks laboratory of my own life.

At about the same time, I worked as a private detective (you heard that right) for an idealistic group of investigators who had been trained by the legendary Hal Lipsett, famous not only for inventing the bug-in-the-martini-olive, but for bringing romance, sophistication, intelligence, and savoir faire to the shadowy world of detectives. In addition to its opportunity to help me find my long-lost father, I was beguiled by the intrigue of detective work and seduced by the left-leaning politics and intellectual chops of this group of detectives. One of the partners, Josiah "Tink" Thompson, gave up a tenured position at Haverford, where he was a renowned Kierkegaard scholar, to become a gumshoe.

Most of my work was undercover, and I used my feminine wiles to gather compromising evidence in cases that invariably aligned with my politics. For my first undercover assignment, I had to gather incriminating evidence on corrupt cops who had planted drugs on our client, all while wearing a wire concealed in my bra.

One night, however, I faced a moral crisis. We'd arranged a dinner party where every single person was undercover except our target. The plan was flawless. For several months, I had cultivated a relationship with the target, and he didn't have the slightest idea what was coming. While I felt confident that he was the "bad guy," he had also become a real person to me, not a character in a TV series or a mystery novel.

Our client was his ex-wife, whose son was set to inherit a great deal of money. Our would-be dinner guest was planning to kidnap the son and abscond with the inheritance. To my young self, who still saw the world as black and white, good and bad, right and wrong, I was on the side of the good, without question. It seemed straightforward. Until it didn't.

As we were preparing the perfect sting, I had an unexpected change of heart. As much as I relished the power and romance of being the protector of the innocent—not to mention the Mata Hari allure and intrigue—something clicked inside, and I couldn't go through with it. It was as if I had awoken from a dream. All of a sudden I realized that the end did not justify the means and the whole subterfuge felt wrong to me. This particular moral principle (does the end ever justify the means?) had come alive for me, out of the armchair and into real life—it was no longer a philosophical debate. There were human beings involved, and they were all complicated. Even this guy, the potential kidnapper, whom I had come to know under false pretenses, who had believed me, who *trusted* me. I just couldn't betray a person whose trust I had cultivated. I was no longer going to sign on for something just because somebody told me it was right and true. I had to taste and feel it for myself.

Integrity Is an Inside Job

I was a perfect candidate, then, for a religion that stated at the outset, don't believe anything just because the Buddha said it was true. Test it and find out for yourself. Sign me up! I was the proverbial grazing animal who needed the biggest possible pasture to wander about in. If I was to be "tamed," I would have to be given the ultimate freedom to discover *for myself* what was right and what was wrong.

My Buddhist practice gave me the tools to experience firsthand what led to happiness and what led to suffering. This involved a nonintellectual seeing. Meditation practice was about looking deeply at the mind and body and *connecting the dots*. If I engage in unskillful behavior, it leads to suffering. If I engage in skillful behavior, it leads to well-being. These are not commandments handed down from on high. They're pragmatic truths that follow the same lawfulness as all of nature.

It seems so straightforward, and yet integrity and morality have been in the aegis of religion and philosophy from time immemorial. The first trades in commandments and the latter in intellectual debates. Neither of these modes worked for me. I rebel against commandments and argue endlessly with intellectual debates.

Remember Cicero's idea about scruples as pebbles in your shoe? The funny and beloved meditation teacher Sylvia Boorstein elaborated on this idea at a silent meditation retreat I sat with her a few years ago. She talked about how settling the mind creates the conditions for all the moral transgressions we've ever committed to appear, like a cork seeking the surface of the water. What she called our personal "scrupulosity" machines. (It turns out *scrupulosity* is actually a psychological condition that's a form of obsessive-compulsive disorder involving excessive attachment to religious rituals and moral codes, but I thought she made up the word at the time and certainly didn't mean it to refer to a compulsion.)

She compared the process to Bessel van der Kolk's descriptions of how our bodies send us messages in his best-selling book *The Body Keeps the Score*. As van der Kolk says, the body does keep score, but so does the mind. Moral transgressions give rise to *non-equanimity* while equanimity, on the other hand, gives rise to atonement and the desire to make things right. We keep the mind and body busy (and sometimes get into a world of trouble) just to avoid the "scrupulosity machine" from revealing all the ways we've missed the mark.

Shame arises as a natural and even beautiful human response

meant to be instructive and help us to survive as social animals. But shame *feels* awful. For me, it's one of the yuckiest emotions, far more challenging than anger, fear, or sadness. If we can't tolerate the feeling of shame (as I couldn't), we seek distraction. Often this behavior (like using drugs, shoplifting, sex, you name it) creates yet more shame. It begins to feel overwhelming, impossible to bear, so we need ever-stronger and more frequent distractions, and so it goes. Meanwhile, the true source of suffering is never addressed.

In my experience, once I let myself feel the shame, things start to release and soften. The judgment I hold against others no longer has a foothold when I acknowledge my own shortcomings. I see that the false front of superiority I hide behind actually serves to separate me from others and increase my suffering.

When not seen clearly, shame is taken personally, reifying the sense of a separate self and keeping us preoccupied with our own shortcomings If we can turn toward the embarrassing and yucky feeling of shame, we can learn from it and it can develop into remorse. From the place of remorse, we can discern if there are any restorative actions we can take. If we meet the shame with judgment, however, it becomes a "near enemy" of shame: guilt. Guilt hardens us to ourselves and to others. It's the flip side of blame. It hardens into fixed positions and views of ourselves and others.

Unable to tolerate our own shame, we project blame and anger toward others that can quickly escalate into violence when unchecked. The whole edifice of blame is predicated on the idea of finding the one damn person who started it all, but modern neuroscientists, like Robert Sapolsky, whom we met in chapter 6, have been poking holes in this idea for years. Their understanding seems to align with the Buddhist idea that there's no consistent and continuous self at the bottom of our actions. Rather, our actions emerge from a complex web of inexorable causes and conditions. It's not necessary to embrace non-self, however, to ease up on the blame game. We can all begin by remembering our shared common humanity: everyone is simply trying to get their needs met in the only ways they know how.

In 2004, I interviewed Marshall Rosenberg for a Buddhist journal called *Inquiring Mind*, because many of my fellow practitioners had become intrigued by his work on Nonviolent Communication (NVC). Although Marshall was not a Buddhist, many of his ideas synced up with Buddhist teachings. In fact, when I asked him about this he said, "I'm from Detroit, so I don't use the word *meditation*. I see it as getting my shit together. This means getting clear on how I choose to live before I go out in the world."

Both Buddhist psychology and NVC agree that the worst way to elicit compliance is to demand it. And the best way is to invite people to see for themselves how the desired behavior could contribute to their own well-being and the well-being of others. Although these principles are basic to child psychology, they still work well on adults like me, expressed cogently in Marshall's own words:

> Demands . . . lead to fear, guilt, and resentment. When people feel coerced or controlled, they are less likely to act with goodwill and more likely to rebel or comply begrudgingly. Genuine cooperation arises from mutual understanding and voluntary agreement.

Guidelines to Live By

As one of the first ten people to be certified to teach MBSR, I was on the front lines of the mindfulness "movement" and had already taught hundreds of people in many different settings by the time the inevitable backlash that accompanies every trend occurred. One of the most strident arguments against MBSR was the supposed omission of explicit teachings on ethics (*sīla* in Pāli). The *sīla* teachings always made sense to me, especially in the context of a retreat or a monastery where people had to agree on guidelines for two important reasons: (1) to live peaceably together, and (2) to create the conditions for awakening.

Typically taught as a set of precepts agreed to before entering the community, these ethical guidelines create a safe container for the hard work of training the mind. MBSR classes did indeed have guidelines that were agreed to in order to participate in the class. They were appropriate for the type of community that was established in an 8-week class and were part of the "invisible" container that created safe conditions for vulnerability and deep inquiry.

As the gadflies railed against the lack of explicit ethical training in MBSR, I saw my students go into ethical overdrive, questioning their own behavior in ways I thought impossible for such a short class. Some resigned from practicing medicine at Kaiser on "moral grounds" or radically rethought their approach to interactions with patients. I saw "scrupulosity machines" busy at work grappling with infidelities, betrayals, and the whole poignant shebang of human foibles.

The kind of moral inquiry that took me weeks of silent, intensive practice was emerging for my students in a class that met only once a week for two and a half hours. This was a stunning testament to the power of mindfulness practice itself to be a driver of ethical behavior. It increased my faith in the simple power of sitting and paying attention to your own mind.

Long before I knew about Cicero's pebbles and scrupulosity machines, I used a similar metaphor in my classes, describing the misdeeds we carry as an invisible backpack weighted down with pebbles (likely rocks in my case, perhaps too many to fit in a shoe). We haul around these invisible backpacks of shortcomings, and they weigh us down and feed a sense of isolation rather than connection. *We alone are flawed and imperfect*, and we have to work ever harder to compensate, resulting in inauthenticity and a vicious cycle of feeling farther and farther away from our true selves and from others. We put ourselves in metaphorical jails where, like the 11.5 million incarcerated individuals around the world, we are removed from the compassion, love, and understanding that are the only remedies for our distress and sense of alienation.

It seemed to me the very point of the practice was to activate our built-in antennae, our moral gyroscopes, sending signals when we go off course. It is the *simple practice of paying attention* in itself that sensitizes us to ethical danger zones and clarifies our values, so they blossom from the inside out.

I can offer no proscriptions about precisely what that means for you. It is between you and you, or you and your God. Your job is to clarify your moral compass, point yourself in that direction and then pony up to the scrupulosity machine in real time so you won't have to pay with interest later. You'll never get it perfectly right, and that really isn't the point. But the more you're able to course correct, the more equanimity will be available to you.

Shared Human Values

However personal this job may be, it turns out that most of us discover we have fundamental values in common. The Dalai Lama is famous for saying "My religion is kindness." It's pretty hard to argue with that.

Kindness and care are hardwired into all social creatures, for good reason. We are utterly dependent on one another from the moment we're born until the moment we die. Without an orientation of care toward one another we would perish.

A movie, *One Life*, came out recently starring Anthony Hopkins that dramatized the life of Sir Nicholas Winton, who managed to save 669 Jewish children from the Holocaust. Despite all odds, and with no acclaim, he used his ingenuity and altruistic motivation to arrange trains to transport the children to England, and to find homes for each of them. In an interesting twist on the Dalai Lama's comment, Winton said, "My religion is ethics." Born to a Jewish family that converted to Christianity, Winton had a "distaste for religion." What motivated him to save these children, at great personal risk and expense, was a sense of moral duty that had nothing to do with either Judaism or Christianity. It was his inner gyroscope flashing *tilt, tilt, tilt.*

So, let's connect the dots here. Whether you get there through religion, or as a private detective, with a God or without a God, to live a life based on compassion, kindness, and generosity is to live an ethical life. If we're well grounded in ethics, we're less moved by the worldly winds. So far, we've used both the gyroscope and the trimming of the sails as metaphors for equanimity. To continue with the boat metaphor, integrity is like the keel of the boat. Our core values provide the weighty ballast that keeps the boat upright in the stormy seas of praise and blame, gain and loss, joy and sorrow. The more closely aligned with our core values, the greater the access we have to equanimous responding.

The four immeasurables—lovingkindness, compassion, sympathetic joy, and equanimity—are inextricably connected to ethics, because if you commit to live a life of kindness and compassion, you're automatically living an ethical life. Likewise, a truly ethical life is a life of equanimity. And for that matter, mindfulness without equanimity and ethics is merely paying attention.

Values Clarification

The values and goals that lie beneath the surface, as would-be drivers of our actions, can get blurred out, or in the language of psychology, become *obstructed*. Catherine Juneau, the Montreal researcher we met back in chapter 4, conducted research on how people with high levels of equanimity can better handle stress, anxiety, and depression—especially when their underlying personal goals or values are obstructed.

Juneau built her work on top of research that showed that living in alignment with one's values enhanced well-being and that by contrast when one's values are obstructed distress results. She offers the example of someone who highly values physical health yet avoids going to the gym due to overwork. Her research shows that people with higher levels of trait equanimity (i.e., it's a part of who they are) are better equipped to handle stress, anxiety, and depression—especially when their personal values or goals are obstructed.

By extension, then, equanimity keeps one's values intact, despite the occasional obstruction. And, by not throwing in the towel on our values, we gain the peace of mind that comes from living in alignment with them, which enhances our equanimity. It's a two-way street. Values strengthen equanimity and equanimity strengthens values.

Remember our friend David Creswell from chapter 4? Always seemingly ahead of the curve, David wrote a paper back in 2005 on the power of values affirmation to buffer both the physical and the psychological adverse effects of stress. The study's findings suggested that reflecting on personal values can indeed *lower the stress response*. Such research would indicate that understanding the values that underly our actions can indeed provide the ballast in the keel of our boat that helps us weather a storm. Without our values, we are easily left at sea. As the Stoic philosopher Seneca said, "There is no favorable wind for the sailor who doesn't know where to go."

The power of short interventions around values clarification was first brought to my attention by Kelly McGonigal, a colleague at Stanford's Center for Compassion and Altruism Research and Education (CCARE). Kelly is a health psychologist who's particularly gifted at making scientific research as scintillating as a screenplay. We worked together for several years on training teachers to become Compassion Cultivation Training instructors. We were a founding faculty of five women, and Kelly encouraged all of us to not only translate research into meaningful tools for everyday life but also to "invent" creative ways to turn research findings into engaging exercises. Kelly was always interested in "whole person" learning and none of us were surprised when she wrote a book called *The Joy of Movement*. As a born teacher (and the daughter of teachers), Kelly understood instinctively that learning is best accomplished by employing the entire network of heart, mind, and body.

At Stanford I also had the opportunity to work with Dr. James R. Doty, the neurosurgeon who founded CCARE. In his latest book, *Mind Magic*, James makes an evidence-based case for how intention can recruit parts of the brain like the salience network—offering a scientific explanation for how intentions manifest in the outer world. The salience network acts as the brain's spotlight, identifying and prioritizing the most important stimuli to focus on. He also suggests that intentions result in well-being only when they're linked to eudemonic goals as opposed to hedonic goals (i.e., prioritizing core values over achievement and material gain).

He draws a welcome distinction from the magical thinking touted in New Age manifestos like *The Secret* or *The Law of Attraction*. Though there are scientific explanations for how the mind can manifest in the outer world, that power will only lead to happiness if it's harnessed in the service of core values rather than personal gain.

A Values Reflection

The following short practice—inspired by the research and by Kelly—will give you a good feel for the power of clarifying values. To begin, read through the list of core values on the next page and see if one of them stands out to you.

From there, you can approach this exercise in a few different ways. You might choose to spend a few minutes writing about this value, including what it means to you, reflecting on people who have inspired this value in you, and recalling times when you have embodied it yourself.

Another option would be to choose one value from the list and decide, just for one day (tomorrow, for example) to align your life a little more closely to that value. For instance, if you choose kindness, set the intention tomorrow morning to be a little kinder for the day and just see what happens. Unleash your salience network on kindness. Notice if this intention reminds you to be more patient with a loved one, or more generous with a friend. Experiment with the principle of "less is more." Don't set the bar too high. Celebrate the small moments.

VALUES

- Authenticity
- Autonomy
- Balance
- Beauty
- Compassion
- Challenge
- Citizenship
- Community
- Contribution
- Creativity
- Curiosity
- Fairness
- Friendship
- Fun
- Growth
- Honesty
- Humor
- Inner Harmony
- Justice
- Kindness
- Learning
- Love
- Loyalty
- Openness
- Peace
- Service
- Trustworthiness

Contemplating ethical questions can seem weighty and judgmental—even scary. For some people it evokes memories of strict religious schools or punitive parents. For people like my younger self, it can provoke instinctive rebellion. Remember back in chapter 5 how Jack Bauer used the metaphor of volume as a way to describe the "quiet ego"? This metaphor allowed him to avoid the polarity of either/or thinking about the ego. He didn't want to suggest we get rid of the ego altogether, just lower the volume. Let's borrow that metaphor here as a way of approaching integrity.

None of us will likely become saints, living lives of impeccable virtue. But we can reduce the frequency of the deviation indicators coming from our moral gyroscope. Each minute shift in the direction of core values turns the volume down on *papañca* and perseveration. Each moment of reckoning lightens the weight of our backpack and leads to compassion and equanimity.

For fourteen years, I taught Compassion Cultivation Training to hundreds of people, which helped me to see that our motivations differ. Some people are compelled by the fact that cultivating compassion and other virtues is directly linked to our personal happiness and well-being. Other people are only energized if there's clear benefit for others. Many of us feel motivated by a combination of the two. Whatever the case, it transcends following rules and commandments. It's about the well-being and equanimity that result.

If you're someone who feels more motivated by the impact that living ethically has on others, you might consider how your commitment to non-harming creates safety for other people. It is a gift not just for you but also for all you come in contact with. As sentient beings, we're all wired to detect threats. In the presence of people who exhibit integrity, our innate threat detectors can begin to relax. As I mentioned, during meditation retreats, we adhere to specific precepts that create a rare atmosphere of safety, relaxation, and ease.

A remarkable example of what happens within a secure safe container is seen in the behavior of animals in the sanctuary world of the Galápagos Islands. While the animals there are indeed exotic, what struck me most was their complete fearlessness around humans. These animals remain unperturbed by the towering, two-legged visitors who faithfully adhere to designated paths and strict protocols, respecting their behavior and habitat.

A world where we live our values and communicate our integrity and respect and care for others is a playground of equanimity.

Epilogue

Drop by Drop

This dewdrop world—
Is a dewdrop world,
And yet, and yet...
—**Kobayashi Issa**

As I said at the outset, I never planned to write a book about equanimity, and I certainly didn't anticipate revealing deeply personal stories or flying to Arizona to get my brain zapped with transcranial ultrasound. Much like my whole cockamamie career, this book had its own ideas.

As the scope of the book grew, I did my best to engage with diverse points of view, listen deeply and openheartedly, and report as accurately as I could. If you happen to be a neuroscientist, religious scholar, or Indigenous leader, you may encounter the limits of my understanding. As Swami Beyondananda wrote in his "Guidelines for Enlightenment": "Everything I have told you is channeled. That way, if you don't like it, it's not my fault."

In fact, I followed the inner whisperings of this book because of my profound conviction in the power of equanimity to restore balance to a world that often feels like it is tipping off its axis. Though I can't know what the state of US or global politics will be by the time this book reaches your hands, it seems reasonable to predict that the flames of reactivity and polarization will continue to be fanned. My deepest hope is that *quiet strength* will offer an alternative to the noisy bullying that has been dominating the political and economic landscape—a kind of power that creates the possibility for compassion, wisdom, peace, and equity to prevail.

Equanimity is a fundamental capacity we all have, to varying degrees. If you picked up this book in the first place, chances are you have more than the average bear. I taught Compassion Cultivation Training to hundreds of people over the years and consistently found that the people who signed up were already super compassionate. They didn't need more compassion *training*; they needed more *permission* to be compassionate (along with a little conceptual clarity). Likewise, choosing to read this book and making it to the end suggests you already have all the equanimity you need. You just need a voice that champions quiet strength when the rest of the world wants you to be enraged, outraged, and off-balance.

One of my first jobs teaching and developing contemplative curricula for a research study involved overweight women. I co-facilitated with a nutritionist who taught us about the different kinds of satiety signals the body sends to the mind. They range from hormonal cues that signal general fullness to taste-specific satiety. Not surprisingly, the hormonal cues can be quite subtle and are easily overridden by visual and olfactory stimulation, causing us to overeat.

Likewise, because equanimity is quiet, subtle, and sometimes amorphous, it can easily get drowned out in the din of noisy egos, marketing campaigns, and clever algorithms designed to appeal to our most primitive instincts. My hope is that this book will offer encouragement, permission, and guideposts back to values that aren't typically touted or easily packaged: equanimity, harmony, complexity, ambivalence, humility, vulnerability.

The impulse toward quiet virtues can also be stifled through the pernicious affliction of self-doubt, as in the following quote attributed to Henry Miller:

> Every day we slaughter our finest impulses. That is why we get a heartache when we read those lines written by the hand of a master and recognize them as our own, as the tender shoots which we stifled because we lacked the faith to believe in our own powers, our own criterion of truth and beauty.

Whether or not you continue to work with some of the many practices offered in this book, your salience network is now primed to consider the possibility of door #3: the equanimous response. Even when things get intense.

Remember, you don't have to *feel less* to be equanimous, you just have to unhook from the grasping and aversion that occlude your innate capacity for wisdom.

Equanimity will never be an Olympic event. You'll never get a medal or a gold star or a big *E* to pin on your chest. Equanimity is just not that into the praise-and-blame thing. It's likely to involve a certain amount of letting go of the addictive habit of mainlining energy through habits like social media or compulsive texting in favor of something quieter, more reliable, and ultimately more nourishing.

At the same time, you don't have to put in ten thousand hours on a meditation cushion to become more equanimous. You don't even have to meditate at all, though it's one of many ways to cultivate equanimity. It might come down to something as simple as pausing for a moment to ask yourself, "What would an equanimous response look like in this moment?"

Imagine if we all did that in any given moment. It would be an "exotic moment," as Pablo Neruda wrote in his famous poem "Keeping Quiet." A moment in which . . .

> Fishermen in the cold sea
> would not harm whales
> and the man gathering salt
> would look at his hurt hands.

Above all, don't lose heart. The possibility of a *future we can love*, as my dear friend Susan Bauer-Wu wrote about in her beautiful book of the same name, is built upon humble and often quiet moments of equanimity, resistance, clarity, and love.

May each of us find a way to bring balance and tenderness to this floating world, moment by moment, drop by drop.

Acknowledgments

Profound gratitude to:

Jon Kabat-Zinn: my first and most important mentor. Thank you for your generosity, love, impeccability, and steadfastness lo these many years and for always believing in me.

Tirzah Firestone: coach, therapist, spiritual guide, oracle. Your love, wisdom, and guidance have buoyed me, strengthened me, and reminded me, again and again, of my deepest and truest intentions for this book.

Barry Boyce: You are a prince, a literary midwife, a book whisperer, a magician, and a wordsmith. I wanted you to write this book for me, but somehow you drew it out of me, all the while shapeshifting from thought partner to editor to coach to dharma buddy to co-conspirator and to movie critic, as needed.

All the many dharma teachers I've had the privilege to learn from and particularly the triumvirate who started me on the path forty-five years ago and have been dharma beacons ever since: Joseph Goldstein, Sharon Salzberg, and Jack Kornfield.

Buddhist teachers and scholars who influenced, expanded, and deepened my understanding of equanimity: Matthew Brensilver, Gil Fronsdal, Mingyur Rinpoche, Shinzen Young, Bhikkhu Anālayo, John Dunne, Sylvia Boorstein, and Nikki Mirghafori.

Ralph Metzner: spiritual guide, friend, and mentor who helped me

ACKNOWLEDGMENTS

heal the trauma you have read about in this book and to envision my true path.

Marshall Rosenberg: communication and relationship genius who taught me the heart of equanimity without ever using the word *equanimity*.

Interviewees: Please see the list of thirty remarkable individuals who engaged wholeheartedly in sharing their unique perspectives on equanimity. These conversations informed, enlivened, inspired, illuminated, and deepened this entire project and were often what kept me going through the long fallow periods of writing this book.

Nancy Rothschild, Wendy Zerin, Pia Stern, and Laurie Graham: dear friends, dharma sisters, cheerleaders, emotional support, and trenchant editorial assistance.

Zara Houshmand, Barbara Bogatin, Susan Bauer-Wu, Cliff Saron, Erika Rosenberg, Joan Kotun, and Rick Hanson: Beta readers extraordinaire. Thank you for your generosity and expertise.

Tom Lane: dear friend, dharma buddy, and brilliant editor who contributed to the proposal and the first few chapters.

Stephanie Tade: literary agent who had a bigger vision for this book than I dared imagine. Thank you to Susan Bauer-Wu for connecting me to such a kind, skillful, and like-minded advocate for equanimity.

Gabriella Page-Fort: executive editor at HarperOne who, like my agent, saw the potential of this book even before I did and offered her full and brilliant support. I could not possibly wish for a better ally in bringing the gift of equanimity to the world. And thanks to the entire team at HarperOne for their contributions.

Wendy Millstine: acquisition editor at New Harbinger who suggested I write another book.

Barbara Gates: writer, editor, and dharma sister who was always receptive to my ideas and the very first to encourage me to write for publication.

Michael, my North Star, and Sofi, my heart that lives outside my body.

List of Interviewees

This book would not be possible without the contributions from the following individuals, who each agreed to be interviewed and to share their unique perspectives on equanimity. For more information, please see their websites below.

Bruce Alderman, M.A., https://appliedmetatheory.org/author/brucealderman/

Jack Bauer, PhD, www.transformativeself.org

Swami Beyondananda (a.k.a. Steve Bhaerman), https://www.wakeuplaughing.com

Tom Block, https://www.tomblock.com

Habib Todd Boerger, PhD, https://www.habibboerger.com/

David Creswell, PhD, https://www.cmu.edu/dietrich/psychology/directory/core-training-faculty/creswell-david.html

Richard Davidson, PhD, https://www.richardjdavidson.com

Andrew Dreitcer, PhD, https://cst.edu/faculty/andrew-dreitcer

John Dunne, PhD, https://www.johnddunne.net

Rabbi Tirzah Firestone, PhD, https://www.tirzahfirestone.com

Gil Fronsdal, PhD, https://www.insightmeditationcenter.org/teachers-and-dharma-leaders

LIST OF INTERVIEWEES

Roshi Joan Halifax, PhD, https://www.joanhalifax.org

Catherine Juneau, PhD, https://www.researchgate.net/profile/Catherine-Juneau

Jon Kabat-Zinn, PhD, https://jonkabat-zinn.com

Kritee Kanko, PhD, https://www.kriteekanko.com

Jack Kornfield, PhD, https://jackkornfield.com

Charles Lawrence, https://danceforallpeople.com/charles-lawrence-interview

Tim Lomas, PhD, https://www.drtimlomas.com

Rhonda V. Magee, JD, https://rhondavmagee.com, https://mountiris.com/

Iris Mauss, PhD, https://psychology.berkeley.edu/people/iris-mauss

Tim Ryan, JD, https://politics.uchicago.edu/fellows/former-fellows/tim-ryan

Matthew Sacchet, PhD, https://meditation.mgh.harvard.edu/sacchet

Sharon Salzberg, https://www.sharonsalzberg.com

Zindel Segal, PhD, https://betterineverysense.com

David R. Vago, PhD, https://contemplativeneurosciences.com

Michael Yellow Bird, MSW, PhD, https://www.indigenousmindfulness.com/about

Rev. Aizaiah Yong, PhD, https://www.spiritedrenewal.org

Shinzen Young, https://www.shinzen.org

Kiliii Yüyan, https://www.kiliii.com

Wendy Zerin, MD, https://www.wendyzerin.com (for information about Kaiut Yoga see: https://kaiutyoga.com)

Notes

Prologue

2   *As a guest on Sharon Salzberg's podcast*: Sharon Salzberg, *Metta Hour with Sharon Salzberg*, Episode 179, "Margaret Cullen," Be Here Now Network, March 22, 2022, Podcast, 68 min., 32 sec, https://beherenownetwork.com/sharon-salzberg-metta-hour-ep-179-margaret-cullen/.

3   *The idea of letting "accidents" happen*: Sharon Salzberg, "The Pretense of Accident: Yearning, Not Gripping, for Happiness," On Being, January 4, 2016, https://onbeing.org/blog/the-pretense-of-accident-yearning-not-gripping-for-happiness/.

4   *Mindfulness-Based Emotional Balance (MBEB)*: Margaret Cullen and Gonzalo Brito Pons, *The Mindfulness-Based Emotional Balance Workbook* (New Harbinger, 2015).

Chapter 1: Why Equanimity?

11   *"Equanimity deepens the poignancy"*: Matthew Brensilver, in-person dharma talk, quoted with permission, date unknown.

15   *In the Visuddhimagga*: Bhadantācariya Buddhaghosa, *The Path of Purification: Visuddhimagga*, trans. Bhikkhu Nanamoli (BPS Pariyatti Editions, 2003).

16   *is the four* brahmavihārās: Gil Fronsdal, "The Four Faces of Love: The Brahma Viharas," Insight Meditation Center, accessed March 24, 2025, https://www.insightmeditationcenter.org/books-articles/the-four-faces-of-love-the-brahma-viharas/.

25   *In neurobiology, when trauma is triggered*: Nikayla Jefferson, "How Buddhism Can Inform Climate Activism: A Conversation with Kritee Kanko," Yale Climate Connections, March 29, 2023, https://yaleclimateconnections.org/2023/03/how-buddhism-can-inform-climate-activism-a-conversation-with-kritee-kanko/.

25   *Buddhist teacher Michael Carroll*: Michael Carroll, "From the Wisdom Seat," *The Wisdom Seat*, May 20, 2025, https://www.thewisdomseat.org/post/june-2025-newsletter.

NOTES

25 *"The future is frightening.":* Caroline Hickman, Elizabeth Marks, Panu Pihkala, Susan Clayton, R. Eric Lewandowski, Elouise E. Mayall, Britt Wray, Catriona Mellor, and Lise van Susteren, "Climate Anxiety in Children and Young People and Their Beliefs About Government Responses to Climate Change: A Global Summary," *Lancet Planetary Health* 5, no. 12 (2021): e863–e873, https://www.sciencedirect.com/science/article/pii/S2542519621002783.

25 *Among respondents aged sixteen to twenty-five:* Roger Harrabin, "Climate Change: Young People Very Worried–Survey," BBC, September 14, 2021, https://www.bbc.com/news/world-58549373.

26 *A study reported in* Psychology Today: Liraz Margalit, "Why We're Addicted to Our Smartphones, but Not Our Tablets," *Psychology Today*, November 21, 2015, https://www.psychologytoday.com/us/blog/behind-online-behavior/201511/why-were-addicted-to-our-smartphones-but-not-our-tablets.

27 *In a study recently published:* K. B. Nandita and Santhosh K. Rajan, "Finding Balance in a Digital World: Equanimity as a Predictor of Nomophobia," *Journal of Human Behavior in the Social Environment*, (2024): 1–11, https://doi.org/10.1080/10911359.2024.2362181.

27 *the movie* The Social Dilemma: Jeff Orlowski, dir., *The Social Dilemma*, Exposure Labs Production, Netflix, 2020, Docudrama, 94 min., https://thesocialdilemma.com/.

27 *said in a recent interview:* Bill Maher, *Real Time with Bill Maher*, Season 21, episode 636, "Tristan Harris, James Kirchick, Matt Duss," Omny Studio, October 13, 2023, podcast, 61 min., https://podcasts.apple.com/us/podcast/ep-636-tristan-harris-james-kirchick-matt-duss/id98746009?i=1000631279516.

27 *Researchers at Yale:* Bill Hathaway, "'Likes' and 'Shares' Teach People to Express More Outrage Online," *YaleNews*, August 13, 2021, https://news.yale.edu/2021/08/13/likes-and-shares-teach-people-express-more-outrage-online.

28 *In* Generations: Jean M. Twenge, *Generations: The Real Differences Between Gen Z, Millennials, Gen X, Boomers, and Silents—and What They Mean for America's Future* (Atria, 2023).

28 *In an interview on NPR:* Michaeleen Doucleff, "The Truth About Teens, Social Media and the Mental Health Crisis," *All Things Considered*, NPR, April 25, 2023, https://www.npr.org/sections/health-shots/2023/04/25/1171773181/social-media-teens-mental-health.

28 *In a* New York Times *opinion piece:* Jonathan Haidt and Jean M. Twenge, "This Is Our Chance to Pull Teenagers Out of the Smartphone Trap," *New York Times*, July 31, 2021, https://www.nytimes.com/2021/07/31/opinion/smartphone-iphone-social-media-isolation.html.

NOTES

28 *Writer and influencer Rebecca Solnit:* Rebecca Solnit, "Is social media just a grand Buddhist scheme to teach me that reactiveness is optional? #thankshaters," Facebook, March 20, 2024, https://www.facebook.com/share/p/1KdnDu8qd8/.

29 *In a series of studies:* Antoine Lutz, Daniel R. McFarlin, David M. Perlman, Tim V. Salomons, and Richard J. Davidson, "Altered Anterior Insula Activation During Anticipation and Experience of Painful Stimuli in Expert Meditators," *NeuroImage* 64 (2013): 538–46, https://doi.org/10.1016/j.neuroimage.2012.09.030.

Chapter 2: The Worldly Winds

31 *Happy is the man:* Seneca, *Letters from a Stoic*, trans. Robin Campbell (Penguin Classics, 1969).

32 *A more poetic phrase:* Dzigar Kongtrul, *Peaceful Heart: The Buddhist Practice of Patience* (Shambhala, 2020); Encyclopedia of Buddhism, "Eight Worldly Concerns," updated March 15, 2025, https://encyclopediaofbuddhism.org/wiki/Eight_worldly_concerns.

33 *There's a frequently retold story:* Thank you to Barbara Gates for reminding me of this story. Barbara Gates, "Is That So?" *Inquiring Mind* 27, no. 1 (Fall 2010), https://inquiringmind.com/article/2701_28_gates.

34 *"Do everything with a mind":* Achaan Chah, *A Still Forest Pool: The Insight Meditation of Achaan Chah*, comp. Jack Kornfield and Paul Breiter (Quest, 2004). Ajahn Chah and Achaan Chah reflect two different transliterations of the Thai honorific used for teachers.

34 *"Death and life":* Marcus Aurelius, *Meditations*, trans. Martin Hammond (Penguin Classics, 2006).

35 *In a Psychology Today article:* Iskra Fileva, "Stoicism as a Fad and a Philosophy," *Psychology Today*, August 4, 2022, https://www.psychologytoday.com/ca/blog/the-philosophers-diaries/202208/stoicism-fad-and-philosophy.

36 *As I began researching:* Thomas Block, "Sufism and Hasidism: Shared Spiritual Tales," *Sophia: The Journal of Traditional Studies* (2011), https://static1.squarespace.com/static/631baee9f78b0f047c664e87/t/64e00767421bf64303d70b1d/1692403567325/Sufism+%2B+Hasidism+Sophia.pdf.

37 *As Tom wrote:* Block, "Sufism and Hasidism."

37 *In describing hishtavut he wrote:* Dovid Sears, *The Path of the Baal Shem Tov: Early Chasidic Teachings and Customs* (Jason Aronson, 1997).

39 *"Behold, after a person is worthy":* Chaim Vital, *Shaarei Kedusha: Gates of Holiness* (Providence Univ. Press, 2007).

39 *"A spiritually classic Jewish concept":* Moshe Gersht, *It's All the Same to Me: A Torah Guide to Inner Peace and Love of Life* (Spirit House, 2021).

NOTES

41 *"The Divine Qualities can"*: For background and meanings of Islamic terms, see Ismaili .net—Heritage F.I.E.L.D. "Dictionary and Encyclopedia of Ismailism Entries," Heritage Society, accessed March 20, 2025, http://heritage.ismaili.net/words_all?apage=J.

43 *When the apostle Paul:* Quotations from the Christian Bible come from the New Revised Standard Version.

44 *"Peace is connected to reconciliation"*: Raimon Panikkar, *Cultural Disarmament: The Way to Peace* (Westminster John Knox Press, 1995).

46 *"A serene spirit accepts pleasure"*: *Bhagavad-Gita: The Song of God*, trans. Swami Prabhavananda and Christopher Isherwood (New American Library, 1951).

46 *"The sage is quiet"*: Chuang Tzu, "Action and Non-Action," trans. Thomas Merton, Poetry Foundation, accessed March 24, 2025, https://www.poetryfoundation.org/poems/46884/action-and-non-action.

46 *"The Great Way is not difficult"*: Ramdass.org, "Ram Dass Reads the 'Third Chinese Patriarch,'" accessed March 20, 2025, https://www.ramdass.org/ram-dass-reads-the-third-chinese-patriarch/.

48 *He defines a* generative (en)closure: Bruce Alderman, Edward Berge, and Layman Pascal, "Generative (En)Closures, Bubbles, and Magic Circles: A Chat About Integral Postmetaphysical Spirituality and Religion," *Integral Review* 15, no. 1 (January 2019): 14–39. https://integral-review.org/issues/vol_15_no_1_alderman_berge_pascal_generative_enclosures.pdf.

Chapter 3: Mindfulness and Equanimity Play Well Together

50 *There are two types:* Gil Fronsdal, in-person dharma talk, quoted with permission, date unknown.

50 *in an article in* Frontiers in Psychology: Juliane Eberth, Peter Sedlmeier, and Thomas Schäfer, "PROMISE: A Model of Insight and Equanimity as the Key Effects of Mindfulness Meditation," *Frontiers in Psychology* 10 (October 21, 2019), https://doi.org/10.3389/fpsyg.2019.02389.

51 *In Buddhist teachings:* "Satipaṭṭhāna Sutta: The Foundations of Mindfulness," trans. Nyanasatta Thera, *Access to Insight (BCBS Edition)*, December 1, 2013, http://www.accesstoinsight.org/tipitaka/mn/mn.010.nysa.html.

53 *"To give your sheep"*: Shunryu Suzuki, *Zen Mind, Beginner's Mind: Informal Talks on Zen Meditation and Practice*, ed. Trudy Dixon (Shambhala, 2020).

55 *Pāli has a wonderfully:* Andrew Olendzki, "What Is Papañca?" *Lion's Roar*, accessed March 20, 2025, https://www.lionsroar.com/what-is-papanca/.

59 *Four Noble Truths:* Chögyam Trungpa, *The Truth of Suffering, and the Path of Liberation* (Shambhala, 2009).

NOTES

61 *"the window of tolerance"*: Daniel J. Siegel, *Aware: The Science and Practice of Presence—the Groundbreaking Meditation Practice* (TarcherPerigee, 2020).

61 *"All of humanity's problems"*: Blaise Pascal, *Pascal's Pensées: A Justification of Christianity That Is a Masterpiece of Religious Philosophy*, trans. W. F. Trotter (E. P. Dutton, 1958).

67 *When Zindel Segal and colleagues*: Peter J. Bieling, L. L. Hawley, R. T. Bloch, K. M. Corcoran, R. D. Levitan, L. T. Young, G. M. MacQueen, and Z. V. Segal, "Treatment-Specific Changes in Decentering Following Mindfulness-Based Cognitive Therapy Versus Antidepressant Medication or Placebo for Prevention of Depressive Relapse," *Journal of Consulting and Clinical Psychology* 80, no. 3 (2012): 365–72, https://doi.org/10.1037/a0027483.

Chapter 4: This Is Your Brain on Equanimity

70 *"Although involved in worldly ways"*: *The Suttanipāta: An Ancient Collection of the Buddha's Discourses Together with Its Commentaries*, trans. Bhikkhu Bodhi (Wisdom, 2017).

71 *While some researchers*: Matthew D. Sacchet and Judson A. Brewer, "Advanced Meditation Alters Consciousness and Our Basic Sense of Self," *Scientific American* 331, no. 1 (July–August 2024): 70, https:doi.org/10.1038/scientificamerican072024-2ceUwYKmYLSkfiDRg1gLC1.

72 *Sonication-Enhanced Mindfulness Acquisition (SEMA) lab*: Joseph L. Sanguinetti, Stuart Hameroff, Ezra E. Smith, Tomokazu Sato, Chris M. Q. Daft, William J. Tyler, and John J. B. Allen, "Transcranial Focused Ultrasound to the Right Prefrontal Cortex Improves Mood and Alters Functional Connectivity in Humans," *Frontiers in Human Neuroscience* 14 (2020). https://doi.org/10.3389/fnhum.2020.00052; Brian Lord, Joseph L. Sanguineti, Lisannette Ruiz, Vladimir Miskovic, Joel Segre, Shinzen Young, Mariea E. Fini, and John J. B. Allen, "Transcranial Focused Ultrasound to the Posterior Cingulate Cortex Modulates Default Mode Network and Subjective Experience: An fMRI Pilot Study," *Frontiers in Human Neuroscience* 18 (June 2024): 1392199, https://doi.org/10.3389/fnhum.2024.1392199; Brian Lord, John J. B. Allen, Shinzen Young, and Joseph L. Sanguinetti, "Enhancing Equanimity with Noninvasive Brain Stimulation: A Novel Framework for Mindfulness Interventions," *Biological Psychiatry: Cognitive Neuroscience and Neuroimaging* 10 no. 4 (2025): 384–92, https://doi.org/10.1016/j.bpsc.2024.12.005.

73 *Brain stimulation has been used*: National Institute of Mental Health, "Brain Stimulation Therapies," accessed March 24, 2025, https://www.nimh.nih.gov/health/topics/brain-stimulation-therapies/brain-stimulation-therapies.

77 *"on a mission to alleviate suffering"*: David Vago, "Presenting My Research to The Dalai Lama," Roundglass Living, accessed March 25, 2025, https://roundglassliving.com/meditation/articles/david-vago.

NOTES

78 *Shinzen Young*: Jeff Warren, "How Understanding Enlightenment Could Change Science," The Consciousness Explorers Club, February 26, 2013, https://cecmeditate.com/inscapes4/.

78 *authors of a consequential paper:* Gaëlle Desbordes, Tim Gard, Elizabeth A. Hoge, Britta K. Hölzel, Catherine Kerr, Sara W. Lazar, Andrew Olendzki, and David R. Vago, "Moving Beyond Mindfulness: Defining Equanimity as an Outcome Measure in Meditation and Contemplative Research," *Mindfulness* 6, (2015): 356–72, https://doi.org/10.1007/s12671-013-0269-8.

78 *Building on a framework:* R. J. Davidson, "Affective Style and Affective Disorders: Perspectives from Affective Neuroscience," *Cognition and Emotion* 12, no. 3 (1998): 307–30, https://doi.org/10.1080/026999398379628.

78 *"more rapid disengagement":* Desbordes, "Moving Beyond Mindfulness."

79 *Groundbreaking research:* Antoine Lutz, Daniel R. McFarlin, David M. Perlman, Tim V. Salomons, and Richard J. Davidson, "Altered Anterior Insula Activation During Anticipation and Experience of Painful Stimuli in Expert Meditators," *NeuroImage* 64 (January 1, 2013): 538–46, https://doi.org/10.1016/j.neuroimage.2012.09.030.

80 *"Mindful of the arising of anger":* Desbordes, "Moving Beyond Mindfulness."

80 *In a related paper:* Resh S. Gupta, Autumn Kujawab, and David R. Vago. "The Neural Chronometry of Threat-Related Attentional Bias: Event-Related Potential (ERP) Evidence for Early and Late Stages of Selective Attentional Processing," *International Journal of Psychophysiology* 146 (December 2019): 20–42, https://doi.org/10.1016/j.ijpsycho.2019.08.006.

81 *David Creswell, research psychologist:* Wendy Hasenkamp, *The Mind & Life Podcast*, Episode 52, "David Creswell—Exploring Equanimity," Mind & Life Institute, April 6, 2023, 62 min., https://mindandlife.podbean.com/e/david-creswell-%e2%80%93-exploring-equanimity/.

81 *His lab has done:* J. David Creswell, "Learning to Accept Discomfort Could Help You Thrive," *Scientific American*, November 10, 2023, https://www.scientificamerican.com/article/learning-to-accept-discomfort-could-help-you-thrive/.

82 *In a 2013 study:* Cecilia Westbrook, John David Creswell, Golnaz Tabibnia, Erica Julson, Hedy Kober, and Hilary A. Tindle, "Mindful Attention Reduces Neural and Self-Reported Cue-Induced Craving in Smokers," *Social Cognitive and Affective Neuroscience* 8, no. 1 (January 2013): 73–84, https://pubmed.ncbi.nlm.nih.gov/22114078/.

83 *Evan Thompson:* Thompson's ideas on scientific materialism are outlined in Adam Frank, Marcelo Gleiser, and Evan Thompson, *The Blind Spot: Why Science Cannot Ignore Human Experience* (MIT Press, 2024).

83 *Michel Bitbol:* Michel Bitbol traces the way scientific models of reality are constructed

by abstracting from direct experience. Michel Bitbol, "Is Consciousness Primary?" *NeuroQuantology* 6, no. 1 (2008).

84 *a sixteen-item assessment tool*: Holly T. Rogers, Alice Shires, and Bruno Cayoun, "Development and Validation of the Equanimity Scale-16," *Mindfulness* 12(2):1–14, https://doi.org/10.1007/s12671-020-01503-6.

84 *Joey Weber and Michelle Lowe*: Joey Weber and Michelle Lowe, "Development and Validation of the Equanimity Barriers Scale (EBS)," *Current Psychology* 40(2): 684–98, https://doi.org/10.1007/s12144-018-9969-5.

85 *The scale that Catherine*: Catherine Juneau, Nicolas Pellerin, Elliott Trives, Matthieu Ricard, Rébecca Shankland, and Michael Dambrun, "Reliability and Validity of an Equanimity Questionnaire: The Two-Factor Equanimity Scale (EQUA-S)," *PeerJ* 8 (2020): e9405, https://doi.org/10.7717/peerj.9405.

87 *Matthew Sacchet, a neuroscientist*: Antoine Lutz, Julie Brefczynski-Lewis, Tom Johnstone, and Richard J. Davidson, "Regulation of the Neural Circuitry of Emotion by Compassion Meditation: Effects of Meditative Expertise," *PLoS ONE* 3, no. 3 (2008): e1897, https://doi.org/10.1371/journal.pone.0001897; Amanda J. Shallcross, Allison S. Troy, Matthew Boland, and Iris B. Mauss, "Let It Be: Accepting Negative Emotional Experiences Predicts Decreased Negative Affect and Depressive Symptoms," *Behaviour Research and Therapy* 48, no. 9 (September 2010): 921–29, https://doi.org/10.1016/j.brat.2010.05.025; Heidi A. Wayment, Jack J. Bauer, and Kateryna Sylaska, "The Quiet Ego Scale: Measuring the Compassionate Self-Identity," *Journal of Happiness Studies* 16 (August 2015): 999–1033, https://doi.org/10.1007/s10902-014-9546-z; Emily C. Willroth, Gerald Young, Maya Tamir, and Iris B. Mauss, "Judging Emotions as Good or Bad: Individual Differences and Associations with Psychological Health," *Emotion* 23, no. 7 (2023): 1876–90, https://doi.org/10.1037/emo0001220.

Chapter 5: The Psychology of a Balanced Mind

88 *"If you set up feeling good"*: Mark Epstein, Facebook, June 11, 2024, https://www.facebook.com/share/p/1YYFrRDXcc/.

89 *Jack (a.k.a. Judith) Halberstam*: Judith Halberstam, *The Queer Art of Failure* (Duke University Press, 2011).

89 *"Toxic positivity is forced"*: Susan David, "The Gift and Power of Emotional Courage," TEDWomen, November 2017, Video, 16 min., 38 sec., https://www.ted.com/talks/susan_david_the_gift_and_power_of_emotional_courage.

89 *In her bestselling 2022 book*: Susan Cain, *Bittersweet: How Sorrow and Longing Make Us Whole* (Crown, 2022).

NOTES

89 *late social critic Barbara Ehrenreich:* Barbara Ehrenreich, *Smile or Die: How Positive Thinking Fooled America and the World* (Granta, 2009).

90 *Vanderbilt University calculates:* Vanderbilt University, "The Vanderbilt Project on Unity & American Democracy," accessed March 25, 2025, https://www.vanderbilt.edu/unity/.

91 *Harvard neuroscientist Jill Bolte Taylor:* Jill Bolte Taylor, *My Stroke of Insight: A Brain Scientist's Personal Journey* (Viking, 2008).

92 *"It is important to acknowledge":* Jainish Patel and Prittesh Patel, "Consequences of Repression of Emotion: Physical Health, Mental Health and General Well Being," *International Journal of Psychotherapy Practice and Research* 1, no. 3 (2019): 16–21, https://doi.org/10.14302/issn.2574-612X.ijpr-18-2564.

92 *In a study published in March 2023:* Emily C. Willroth, Gerald Young, Maya Tamir, and Iris B. Mauss, "Judging Emotions as Good or Bad: Individual Differences and Associations with Psychological Health," *Emotion* 23, no. 7 (2023): 1876–90, https://doi.org/10.1037/emo0001220.

93 *A joint study:* Kevin C. Stanek and Deniz S. Ones, "Meta-Analytic Relations Between Personality and Cognitive Ability," *PNAS* 120, no. 23 (2023): e2212794120, https://doi.org/10.1073/pnas.2212794120.

93 *A study by the Harvard School of Public Health:* Benjamin P. Chapman, Kevin Fiscella, Ichiro Kawachi, Paul Duberstein, and Peter Muennig, "Emotion Suppression and Mortality Risk over a Twelve-Year Follow-Up," *Journal of Psychosomatic Research* 75, no. 4 (October 2013): 381–85, 2013 Oct;75(4):381–5. https://doi.org/10.1016/j.jpsychores.2013.07.014.

93 *A 2019 literature review:* Patel and Patel, "Consequences of Repression of Emotion."

94 *I recently listened to an interview:* Adyashanti, "Awareness Explorers Interviews Adyashanti, Part 2—Awakening the Head, Heart and Gut," interview by Jonathan Robinson, *Awareness Explorers*, Awaken, September 14, 2024, Video, 29 min., 59 sec., https://awaken.com/2024/09/awareness-explorers-interviews-adyashanti-pt-2-awakening-the-head-heart-and-gut/.

94 *Trevor Noah:* David Paul Meyer, dir., *Trevor Noah: Where Was I*, Netflix, 2023, comedy special, 68 min.

95 *When I came across the paper:* Willroth, "Judging Emotions as Good or Bad."

96 *her research shows:* Brett Q. Ford, Phoebe Lam, Oliver P. John, and Iris B. Mauss, "The Psychological Health Benefits of Accepting Negative Emotions and Thoughts: Laboratory, Diary, and Longitudinal Evidence," *Journal of Personality and Social Psychology* 115, no. 6 (2018): 1075–92, https://doi.org/10.1037/pspp0000157.

96 *Maya Tamir and her colleagues:* Maya Tamir, Christopher Mitchell, and James J. Gross,

"Hedonic and Instrumental Motives in Anger Regulation," *Psychological Science* 19, no. 4 (2008): 324–28, https://doi.org/10.1111/j.1467-9280.2008.02088.x.

97 *Tim Lomas:* Tim Lomas, Trudi Edginton, Tina Cartwright, and Damien Ridge, "Cultivating Equanimity Through Mindfulness Meditation: A Mixed Methods Enquiry into the Development of Decentering Capabilities in Men," *International Journal of Wellbeing* 5, no. 3 (2015): 88–106, https://doi.org/10.5502/ijw.v5i3.7.

98 *Positive Psychology 2.0:* Paul T. P. Wong, "Positive Psychology 2.0: Towards a Balanced Interactive Model of the Good Life," *Canadian Psychology* 52, no. 2 (2011): 69–81, https://doi.org/10.1037/a0022511.

100 *Jack Bauer:* Jack J. Bauer, *The Transformative Self: Personal Growth, Narrative Identity, and the Good Life* (Oxford University Press, 2021).

100 *Along with his colleague Heidi Wayment:* Jack J. Bauer and Kiersten J. Weatherbie, "The Quiet Ego and Human Flourishing," *Journal of Happiness Studies* 24 (2023): 2499–530, https://doi.org/10.1007/s10902-023-00689-5.

101 *edited volume came:* Heidi A. Wayment and Jack J. Bauer, eds., *Transcending Self-Interest: Psychological Explorations of the Quiet Ego* (American Psychological Association, 2008).

103 *Seven Grandfather Teaching:* Nottawaseppi Huron Band of the Potawatomi, "Seven Grandfather Teachings," accessed August 8, 2025, https://nhbp-nsn.gov/seven-grandfather-teachings/.

103 *One of the key distinctions:* Catherine Juneau, Rebecca Shankland, and Michaël Dambrun, "Trait and State Equanimity: The Effect of Mindfulness-Based Meditation Practice," *Mindfulness* 11 (2020): 1802–12, https://doi.org/10.1007/s12671-020-01397-4.

Introduction to Part II

109 *"In equanimity":* Sadhguru, "Daily Quote March 16, 2023," March 16, 2023, Isha Foundation, https://isha.sadhguru.org/en/wisdom/quotes/date/march-16-2023.

111 *"My way of teaching":* Achaan Chah, *A Still Forest Pool: The Insight Meditation of Achaan Chah,* comp. Jack Kornfield and Paul Breiter (Quest, 2004).

112 *research shows that writing:* Angélica M. Silva and Roberto Limongi, "Writing to Learn Increases Long-Term Memory Consolidation: A Mental-Chronometry and Computational-Modeling Study of 'Epistemic Writing,'" *Journal of Writing Research* 11, no. 1 (2019): 211–43, https://doi.org/10.17239/jowr-2019.11.01.07.

113 *In an article in* Tricycle *magazine:* Gil Fronsdal and Sayadaw U. Pandita, "A Perfect Balance: Cultivating Equanimity with Gil Fronsdal and Sayadaw U. Pandita," *Tricycle* (Winter 2005): https://tricycle.org/magazine/cultivate-equanimity/.

NOTES

Chapter 6: Shifting Perspective

115 *"Since my house burned down"*: Widely attributed to Mizuta Masahide (1657–1723), Japanese samurai, poet, and physician. Mary Yukari Waters, *The Laws of Evening: Stories* (Scribner, 2003).

116 *The following three exercises:* Joseph Goldstein, "Equanimity and Compassion in Challenging Times," Dharma talk at Spirit Rock retreat, July 26, 2018, audio, 61 min., 53 sec., https://dharmaseed.org/talks/51314/.

116 *One of my favorite children's books:* Margery Cuyler, *That's Good! That's Bad!* (Scholastic, 1992).

120 *"Rabbi Yochanan taught":* Dovid Sears, *The Path of the Baal Shem Tov: Early Chasidic Teachings and Customs* (Jason Aronson, 1997).

120 *concept of* fana: For the meaning of the Sufi principle of *fana*, see Mi Ainsel, "Road to Sufism—Baqaa and Fana," Medium.com, August 17, 2023, https://medium.com/@miainsel2/road-to-sufism-baqaa-and-fana-9c29d695593b.

120 *"Don't become united":* Alireza Nurbakhsh, "The Illusion of Self: A Discourse," *Sufi Journal of Mystical Philosophy & Practice* 85 (Summer 2013).

121 *Stanford biologist and neuroscientist:* Robert M. Sapolsky, *Determined: A Science of Life Without Free Will* (Penguin Press, 2023).

122 *The cosmologist Carl Sagan:* Carl Sagan, *Pale Blue Dot: A Vision of the Human Future in Space* (Random House, 1994).

122 *Dacher Keltner's lab:* Xinyu Pan, Tonglin Jiang, Wenying Yuan, Chenxiao Hao, Yang Bai, and Dacher Keltner, "A Balanced Mind: Awe Fosters Equanimity via Temporal Distancing," *Journal of Personality and Social Psychology* 127, no. 6 (2024): 1127–45, https://doi.org/10.1037/pspa0000410.

122 *read Dacher Keltner's book:* Dacher Keltner, *Awe: The New Science of Everyday Wonder and How It Can Transform Your Life* (Penguin Random House, 2022).

123 *The astronaut Ron Garan:* Ron Garan, "The Orbital Perspective," TEDx Douglas, February 10, 2015, video, 17 min., 7 sec., https://www.youtube.com/watch?v=kjqHhrtwHzo.

127 *Ekman says:* Paul Ekman, *Emotions Revealed: Recognizing Faces and Feelings to Improve Communication and Emotional Life* (Henry Holt, 2007).

Chapter 7: Weathering Storms

128 *"The birds have vanished":* Li Po, "Zazen on Ching-t'ing Mountain," in *Crossing the Yellow River: Three Hundred Poems from the Chinese*, trans. by Sam Hamill (BOA Editions, 2000).

128 *made famous by Jon Kabat-Zinn:* Jon Kabat-Zinn, *Wherever You Go, There You Are* (Hyperion, 1994).

NOTES

Chapter 8: Just Like Me

132 *"The first practice":* Lao Tzu, *Hua Hu Ching: The Unknown Teachings of Lao Tzu,* trans. Brian Walker (HarperOne, 1994).

133 *Rosemary Wells wrote:* Rosemary Wells, *Yoko* (Disney/Hyperion, 2009).

134 *As Hannah Arendt confirmed:* Hannah Arendt, "Eichmann in Jerusalem—I," *New Yorker,* February 8, 1963; Hannah Arendt, "Eichmann in Jerusalem—II," *New Yorker,* February 16, 1963.

134 *This terrifying possibility:* Jonathan Glazer, dir., *The Zone of Interest,* Film4, 2023, drama, 105 min.

135 *The bodhisattva vow:* Encyclopedia of Buddhism, "Bodhisattva Vow," accessed March 27, 2025, https://encyclopediaofbuddhism.org/wiki/Bodhisattva_vow.

136 *"Look at my two arms":* Thich Nhat Hahn, "Equanimity: Wisdom of Non-discrimination," video, posted September 8, 2021, by Plum Village App, YouTube, 12 min., 33 sec., https://www.youtube.com/watch?v=6F0xorz5EZY.

137 *Othering and bias:* Eran Halperin, "Psychology of Intergroup Conflict and Reconciliation Lab," Hebrew University of Jerusalem, accessed March 25, 2025, https://www.eranhalperin.com/.

137 *This practice may help:* Thank you to Elissa Epel for inspiration for this guided meditation, "Just Like Me: Exploring Our Common Humanity, the Personal and Universal Mind in the Anthropocene," guided meditation adapted for UC Climate Course, https://www.climateresilience.online/course.

Chapter 9: How to Love and Care Without Attachment

146 *"There is only the trying":* T. S. Eliot, *Four Quartets* (Harcourt, 1943).

146 *There's a beautiful metaphor:* "Sallatha Sutta: The Arrow," trans. Thanissaro Bhikkhu, *Access to Insight (BCBS Edition),* November 30, 2013, https://www.accesstoinsight.org/tipitaka/sn/sn36/sn36.006.than.html.

150 *a simple posture:* Andrew Bein, *The Zen of Helping: Spiritual Principles for Mindful and Open-Hearted Practice* (Wiley, 2007).

151 *shift to a practice of self-compassion:* For excellent, free, self-compassion practices, please visit the Mindful Self Compassion website: Center for Mindful Self-Compassion, "Meditations & Practices," accessed March 24, 2025, https://centerformsc.org/free-meditations-practices.

156 *a charming novel:* Aimee Bender, *The Particular Sadness of Lemon Cake* (Anchor Books, 2011).

156 *Marshall offers:* Marshall Rosenberg, *Nonviolent Communication: A Language of Life,* 3rd ed. (PuddleDancer Press, 2015).

NOTES

Chapter 10: Stepping Stones to Equanimity

158 *"A day without sunshine":* Epigraph is from a 1970s Steve Martin stand-up routine.

159 *"You only want":* "Sayadaw U Tejaniya," accessed March 26, 2025, https://ashintejaniya.org/.

164 *"The only thing":* Barack Obama, "Remarks by the President in Final Press Conference," January 18, 2017, James S. Brady Press Briefing Room, transcript, https://obamawhitehouse.archives.gov/the-press-office/2017/01/18/remarks-president-final-press-conference.

Chapter 11: Bottom-Up Equanimity

165 *"Mr. Duffy lived":* James Joyce, *Dubliners*, ed. Terence Brown (Penguin Classics, 1992).

165 *"I think, therefore I am":* René Descartes, *Discourse on Method*, trans. Laurence J. Lafleur (Collier Macmillan, 1986).

165 *"hard problem" for science:* John Horgan, "How Dave Chalmers Invented the 'Hard Problem,'" John Horgan (The Science Writer), June 27, 2023, https://johnhorgan.org/cross-check/how-dave-chalmers-invented-the-hard-problem; Andy Karr, *Into the Mirror: A Buddhist Journey Through Mind, Matter, and the Nature of Reality* (Shambhala, 2023).

165 *William Beaumont hinted at it:* Charles Stewart Roberts, "William Beaumont, the Man and the Opportunity," in *Clinical Methods: The History, Physical, and Laboratory Examinations*, 3rd ed., ed. H. Kenneth Walker, W. Dallas Hall, and J. Willis Hurst. (Butterworths, 1990).

165 *Michael Gershon:* Michael D. Gershon, *The Second Brain: A Groundbreaking New Understanding of Nervous Disorders of the Stomach and Intestine* (HarperPerennial, 1999).

166 *bottom-up approaches:* Syd Hiskey and Neil E. Clapton, "Enhancing Therapist Courage: Feasibility and Changes in Distress Tolerance and Equanimity Following Martial Arts–Based Radically Embodied Compassion Workshops," *OBM Integrative and Complementary Medicine* 9, no. 2 (2024): 29, https://doi.org/10.21926/obm.icm.2402029.

168 *I spoke with Zindel:* Norman Farb and Zindel Segal, *Better in Every Sense: How the New Science of Sensation Can Help You Reclaim Your Life* (Little, Brown Spark, 2024).

Chapter 12: Uplifting Stories

179 *The shortest distance:* Anthony De Mello, "Myths," in *One Minute Wisdom* (Image, 1985).

180 *Equanimity, as a station:* As Abraham Halkin of the Jewish Theological Seminary notes, Maimonides related the tale of the Sufi adept in a letter to a fellow rabbi, referencing Ibrahim Ibn Adham, an eighth-century Sufi mystic of princely birth. The story would be taken up by later Jewish thinkers to represent a new Jewish mystical station in Thomas Block, "Sufism

and Hasidism: The (Shared) Tales They Tell," *Sophia: The Journal of Traditional Studies* 16, no. 2 (2011): 103–26.

183 *"There is a giant"*: Dmitri Nagishkin, *Folktales of the Amur: Stories from the Russian Far East* (Harry N. Abrams, 1980).

Chapter 13: Breaking the Spell

188 *"Life is a tragedy"*: Charlie Chaplin, "Life is a tragedy when seen in close-up, but a comedy in long-shot," accessed March 26, 2025, https://www.charliechaplin.com/en/quotes/11.

190 *UC Berkeley psychologist Dacher Keltner*: Dacher Keltner, Lisa Capps, Ann Kring, Randall Young, and Erin Heerey, "Just Teasing: A Conceptual Analysis and Empirical Review," *Psychological Bulletin* 127, no. 2 (2001): 229–48, https://doi.org/10.1037/0033-2909.127.2.229.

193 *Jay Leno's funny headlines*: "Jay Leno Best of Headlines Part 17," compiled by Richie Majaw, video compilation, posted May 4, 2020, YouTube, 15 min., 14 sec., https://www.youtube.com/watch?v=lVWpDxOe7gk.

Chapter 14: The Serenity Prayer

195 *One of the most profound*: Reinhold Niebuhr, "The Serenity Prayer," in *The Essential Reinhold Niebuhr: Selected Essays and Addresses*, ed. Robert McAfee Brown (Yale Univ. Press, 1987).

Chapter 15: Taking Refuge, Finding Equanimity

196 *There are two means of refuge*: Widely attributed to Albert Schweitzer, source unknown.

203 *Ruth King's way of framing*: Many thanks to Ruth King for identifying the Three Marks of Existence as Not Perfect, Not Permanent, Not Personal. Ruth King, *Mindful of Race: Transforming Racism from the Inside Out* (Sounds True, 2018).

206 *"Home is the place"*: Robert Frost, "The Death of the Hired Man," Poetry Foundation, accessed March 27, 2025, https://www.poetryfoundation.org/poems/44261/the-death-of-the-hired-man.

206 *epidemic of loneliness*: Aspen Ideas, "Surgeon General's Orders: 3 Tips for Rebuilding Social Connection," accessed March 27, 2025, https://www.aspenideas.org/articles/surgeon-general-s-orders-3-tips-for-rebuilding-social-connection.

Chapter 16: Equanimity in a World on Fire

215 *"My companion in its brightest month"*: Joseph Goldstein, "Venus in the Western Sky," unpublished.

216 *Amaro calls this "unentangled participation"*: Ajahn Amaro, "Unentangled Participation," talk given at Insight Meditation Society retreat, May 4, 2024.

NOTES

216 *Gandhi called it* satyagraha: H. S. L. Polak, "Satyagraha and Its Origin in South Africa." Mahatma Gandhi, accessed March 26, 2025, https://www.mkgandhi.org/articles/satyagraha2.php.

216 *Interestingly, one of the words we use:* Grace Tierney, "The Origin of Unscrupulous, Thanks to a Stone in Your Shoe," Wordfoolery, accessed March 27, 2025, https://wordfoolery.wordpress.com/2021/06/14/the-origin-of-unscrupulous-thanks-to-a-stone-in-your-shoe/.

218 *One stunning exemplar:* Maria Yagoda and Stephanie Sengwe, "Rosa Parks Was Born 112 Years Ago Today: Relive the Civil Rights Activist's Inspiring Moments," People.com, February 4, 2025, https://people.com/politics/rosa-parks-civil-rights-photos/.

219 *In her book:* Rosa Parks and Gregory Reed, *Quiet Strength: The Faith, the Hope, and the Heart of a Woman Who Changed a Nation* (Zondervan, 1994).

219 *work of Reverend James Lawson:* James Lawson, *Revolutionary Nonviolence: Organizing for Freedom* (Univ. of California Press, 2022).

219 *"How may we work":* Howard Thurman, *Deep Is the Hunger* (Friends United Press, 1978).

223 *"make a way out of no way":* The third floor of the National Museum of African American History and Culture has a large exhibit on this principle. National Museum of African American History and Culture, "Making a Way Out of No Way," Smithsonian, accessed March 25, 2025, https://nmaahc.si.edu/explore/exhibitions/making-way-out-no-way.

223 *In her own life:* Rhonda Magee, *The Inner Work of Racial Justice: Healing Ourselves and Transforming Our Communities Through Mindfulness* (TarcherPerigee, 2019).

225 *In 2012, he put out:* Tim Ryan, *A Mindful Nation: How a Simple Practice Can Help Us Reduce Stress, Improve Performance, and Recapture the American Spirit* (Hay House, 2012).

230 *"We don't want to hear":* Kareem Ghandour, "Introduction," *Gaza: Calling for a Dharma Response*, April 27, 2024, https://alokavihara.org/wp-content/uploads/2024/05/Gaza-Calling-for-a-Dharma-Response-updated-april-27-2024.pdf.

231 *"I believe that we":* Joan Halifax, *Standing at the Edge: Finding Freedom Where Fear and Courage Meet* (Flatiron, 2018).

232 *"Joy doesn't betray":* Rebecca Solnit, *Hope in the Dark: Untold Histories, Wild Possibilities* (Haymarket, 2016).

233 *"The sharing of joy":* Keguro Macharia, "Difference: An Audre Lorde Archive," *New Inquiry*, July 24, 2017, https://thenewinquiry.com/blog/difference-an-audre-lorde-archive/; Audre Lorde, "Uses of the Erotic: The Erotic as Power," in *Sister Outsider: Essays and Speeches* (Crossing Press, 1994).

234 *Ram Dass and Paul Gorman:* Ram Dass and Paul Gorman, *How Can I Help?: Stories and Reflections on Service* (Knopf, 1985).

234 *"Some of us were":* Dass and Gorman, *How Can I Help*, 54.

235 *Muste stood silently:* AJ Muste Foundation for Peace + Justice, "A.J. Muste's Life of Activism," accessed March 27, 2025, https://ajmuste.org/aj_mustes-life-of-activism.

236 *"There is a pervasive form":* Thomas Merton, *Conjectures of a Guilty Bystander* (Image, 1968).

Chapter 17: Paradox, Polarization, and ... Uncertainty

237 *"Some of us suffer":* Epigraph repeated widely by Swami Beyondananda and quoted with his permission.

237 *"The bad news is":* While widely attributed to Trungpa Rinpoche, including by many of his students, no source has been verified.

238 *Research from the Max Planck Institute:* Archy O. de Berker, Robb B. Rutledge, Christoph Mathys, Louise Marshall, Gemma F. Cross, Raymond K. Dolan, and Sven Bestmann, "Computations of Uncertainty Mediate Acute Stress Responses in Humans," *Nature Communications* 7 (2016): 10996, https://doi.org/10.1038/ncomms10996.

238 *Pema Chödrön's book:* Pema Chödrön, *Comfortable with Uncertainty: 108 Teachings on Cultivating Fearlessness and Compassion,* ed. Emily Hilburn Sell (Shambhala, 2008).

238 *Zen teacher and psychoanalyst Hubert Benoit:* Hubert Benoit, *The Supreme Doctrine: Psychological Studies in Zen Thought,* 2nd ed. (Liverpool Univ. Press, 1998).

238 *"Enlightenment is simply the willingness":* Charlotte Joko Beck, *Nothing Special: Living Zen,* ed. Steve Smith (HarperOne, 1993).

239 *"Certainty isn't an indication":* Joseph Goldstein with Dan Harris, host, *10% Happier with Dan Harris,* "A Buddhist Antidote to Fear and Anxiety | Devin Berry," Wondery, July 24, 2024, podcast, 68 min., https://podcasts.apple.com/us/podcast/a-buddhist-antidote-to-fear-and-anxiety-devin-berry/id1087147821?i=1000663000800.

239 *Werner Heisenberg formulated:* Caltech, "What Is the Uncertainty Principle and Why Is It Important?" accessed March 27, 2025, https://scienceexchange.caltech.edu/topics/quantum-science-explained/uncertainty-principle.

241 *Naomi Rothman:* Naomi B. Rothman and Shimul Melwani, "Feeling Mixed, Ambivalent, and in Flux: The Social Functions of Emotional Complexity for Leaders," *Academy of Management Review* 42, no. 2 (2016), https://doi.org/10.5465/amr.2014.0355; Laura Rees, Naomi B. Rothman, Reuven Lehavy, and Jeffrey Sanchez-Burks, "The Ambivalent Mind Can Be a Wise Mind: Emotional Ambivalence Increases Judgment Accuracy," *Journal of Experimental Social Psychology* 49, no. 3 (May 2013): 360–67, https://doi.org/10.1016/j.jesp.2012.12.017; Naomi B. Rothman and Gregory B. Northcraft, "Unlocking Integrative Potential: Expressed Emotional Ambivalence and Negotiation Outcomes," *Organizational Behavior and Human Decision Processes* 126 (January 2015): 65–76, https://doi.org/10.1016/j.obhdp.2014.10.005; Naomi B. Rothman, Brianna Barker Caza, Shimul Melwani, and Kate

Walsh, "Embracing the Power of Ambivalence," *Harvard Business Review*, September 14, 2021, https://hbr.org/2021/09/embracing-the-power-of-ambivalence.

241 *Bernie Glassman's three tenets:* Upaya Zen Center, "Practicing the Three Tenets and GRACE in Our Imperiled World," March 15, 2021, https://www.upaya.org/2021/03/practicing-the-three-tenets-and-grace-in-our-imperiled-world/.

242 *A beautiful example of a leader:* Subhash Mehta, "Bhoodan-Gramdan Movement—50 Years: A Review," mkgandhi.org, accessed March 27, 2025, https://www.mkgandhi.org/vinoba/bhoodan.php.

244 *Niels Bohr:* "Niels Bohr 1885–1962: Danish Physicist," in *Oxford Essential Quotations*, ed. Susan Ratcliffe, 5th ed., published 2017, https://www.oxfordreference.com/display/10.1093/acref/9780191843730.001.0001/q-oro-ed5-00001812.

247 *Ella Miron-Spektor:* Ella Miron-Spektor, Kyle J. Emich, Linda Argote, and Wendy K. Smith, "Conceiving Opposites Together: Cultivating Paradoxical Frames and Epistemic Motivation Fosters Team Creativity," *Organizational Behavior and Human Decision Processes* 171 (July 2022): 104153, https://doi.org/10.1016/j.obhdp.2022.104153.

248 *groundbreaking work of Marsha Linehan:* Michael R. Tom and David R. Vago, "Equanimity," in *The Virtues in Psychiatric Practice*, ed. John R. Peteet (Oxford Univ. Press, 2022).

251 *podcast interview with Sam Harris:* Sam Harris, host, *The Making Sense Podcast*, episode 381, "Delusions, Right and Left: A Conversation with 'Destiny' (Steven Bonnell)," samharris.org, August 26, 2024, podcast, 46 min., 46 sec., https://www.samharris.org/podcasts/making-sense-episodes/381-delusions-right-and-left.

253 *Jamil Zaki:* Dan Harris, host, *10% Happier with Dan Harris*, "How and Why to Avoid the Siren Call of Cynicism | Dr. Jamil Zaki," September 9, 2024, video, 77 min., 38 sec., YouTube, https://www.youtube.com/watch?v=9ocHZKbcXu0.

253 *research shows that:* Matthew S. Levendusky and Neil Malhotra, "(Mis)perceptions of Partisan Polarization in the American Public," *Public Opinion Quarterly* 80, no. S1 (2016): 378–91, https://doi.org/10.1093/poq/nfv045.

253 *in his lab they've found:* Yuan Chang Leong, Janice Chen, Robb Willer, and Jamil Zaki, "Conservative and Liberal Attitudes Drive Polarized Neural Responses to Political Content," *Proceedings of the National Academy of Sciences* 117, no. 44 (2020): 27731, https://doi.org/10.1073/pnas.2008530117.

Chapter 18: Brokenhearted Equanimity

255 *"Overcome any bitterness":* As quoted in *The Art of Forgiveness, Lovingkindness, and Peace*, ed. Jack Kornfield (Bantam Books, 2002), 89.

256 *Frank Ostaseski:* Videos of Frank Ostaseski teaching on "The Four Flavors of Love": *Four Fla-*

NOTES

vors of Love: Metta Session *(2023)*, Upaya Zen Center, accessed January 2, 2024, https://www.upaya.org/video/four-flavors-of-love-metta-session-2023/; *Four Flavors of Love: Karuna Session (2023)*, Upaya Zen Center, accessed January 2, 2024, https://www.upaya.org/video/four-flavors-of-love-karuna-session-2023/; *Four Flavors of Love: Mudita Session (2023)*, Upaya Zen Center, accessed January 2, 2024, https://www.upaya.org/video/four-flavors-of-love-mudita-session-2023/; *Four Flavors of Love: Upekkha Session (2023)*, Upaya Zen Center, accessed January 2, 2024, https://www.upaya.org/video/four-flavors-of-love-mudita-session-2023-2/.

256 *When love meets helplessness:* Equanimity as "love meeting helplessness" was first introduced to me by Buddhist teacher Matthew Brensilver.

258 *Zaki's brain-imaging data:* J. Zaki, G. López, and J. P. Mitchell, "Activity in Ventromedial Prefrontal Cortex Co-Varies with Revealed Social Preferences: Evidence for Person-Invariant Value," *Social Cognitive and Affective Neuroscience* 9, no. 4: 464–69, https://doi.org/10.1093/scan/nst005.

259 *In his book* Born to Be Good: Dacher Keltner, *Born to Be Good: The Science of a Meaningful Life* (W. W. Norton, 2009).

259 *In 2014, neuroscientists Tania Singer:* Tania Singer and Olga M. Klimecki, "Empathy and Compassion," *Current Biology* 24, no. 18 (2014): PR875–78, https://doi.org/10.1016/j.cub.2014.06.054.

260 *Daryl and his colleagues studied:* C. D. Cameron and B. K. Payne, "Escaping Affect: How Motivated Emotion Regulation Creates Insensitivity to Mass Suffering," *Journal of Personality and Social Psychology* 100 (2011): 1–15, https://doi.org/10.1037/a0021643.

260 *Paul Slovic originally coined:* Paul Slovic, "'If I Look at the Mass I Will Never Act': Psychic Numbing and Genocide," *Judgment and Decision Making* 2, no. 2 (April 2007): 79–95, https://doi.org/10.1017/S1930297500000061.

261 *Daniel Västfjäll's research:* Daniel Västfjäll, Paul Slovic, and Marcus Mayorga, "Pseudoinefficacy: Negative Feelings from Children Who Cannot Be Helped Reduce Warm Glow for Children Who Can Be Helped," *Frontiers in Psychology* 6 (May 2015), https://doi.org/10.3389/fpsyg.2015.00616.

262 *"According to our theory":* Cameron and Payne, "Escaping Affect."

264 *our shared common humanity:* Jinyoung Park and Christine Lathren, "The Perception of Suffering and Common Humanity in Israel," *OSF* (June 20, 2024), https://doi.org/10.17605/OSF.IO/MFKD9.

Chapter 19: Connecting the Dots: Integrity and Equanimity

273 *"Come, come, whoever you are":* Jalal al-Din Rumi, *Crazy As We Are*, trans. Nevit O. Ergin from Golpinarli's 1992 Turkish translation (Hohm Press, 2015). This is one of the

NOTES

most frequently quoted poems attributed to Rumi, but is not authenticated as his (and it is also not in the earliest manuscripts of the quatrains attributed to him). It is found in the same form in the quatrains of Bâbâ Afzaluddîn Kâshânî (died 1274; Rumi died 1273) and is related to a similar quatrain attributed to Abu Sa'îd ibn Abu'l-Khayr, died 1048 (see page 4 of *Nobody, Son of Nobody: Poems of Shaikh Abu-Saeed Abil Kheir*, renditions by Vraje Abramian [Hohm, 2001]). It is one of the most frequently quoted poems by Turkish Mevlevis (the "Whirling Dervishes") themselves (who have long assumed it to be a Rumi poem), from a Turkish translation of the original Persian.

277 *Hal Lipsett:* Myrna Oliver, "Harold 'Hal' Lipset; Detective Known as the 'Private Ear,'" *Los Angeles Times*, December 10, 1997.

277 *Josiah "Tink" Thompson:* Calvin Trillin, "Tink," *New Yorker*, November 19, 1978.

279 *Bessel van der Kolk's descriptions:* Bessel A. van der Kolk, *The Body Keeps the Score: Brain, Mind, and Body in the Healing of Trauma* (Viking, 2014).

281 *I interviewed Marshall Rosenberg:* Marshall Rosenberg, "Interview with Marshall Rosenberg: The Traveling Peacemaker," interview by Margaret Cullen and Ronna Kabatznick, *Inquiring Mind* 21, no. 1 (Fall 2004), https://inquiringmind.com/article/2101_4w_rosenberg-interview-with-marshall-rosenberg-the-traveling-peacemaker/.

284 *A movie,* One Life: James Hawes, dir., *One Life*, Bleecker Street, Warner Bros. Pictures, 2024, drama, 110 min.

285 *high levels of equanimity:* Aidan Smyth, Catherine Juneau, Seonwoo Hong, Michael John Ilagan, and Bärbel Knäuper, "Facing Obstacles with Equanimity: Trait Equanimity Attenuates the Positive Relations Between Values Obstruction and Symptoms of Depression, Anxiety, and Stress," *Mindfulness* 15, no. 4 (2024): 945–57, https://doi.org/10.1007/s12671-024-02338-1.

285 *David wrote a paper:* J. David Creswell, Shelley E. Taylor, Shane J. Stanton, William T. McEwen, and Bruce S. Sherman, "Affirmation of Personal Values Buffers Neuroendocrine and Psychological Stress Responses," *Psychological Science* 16, no. 11 (2005): 846–51, https://doi.org/10.1111/j.1467-9280.2005.01624.x.

286 *Kelly was always interested:* Kelly McGonigal, *The Joy of Movement: How Exercise Helps Us Find Happiness, Hope, Connection, and Courage* (Avery, 2019).

286 *In his latest book:* James R. Doty, *Mind Magic: The Neuroscience of Manifestation and How It Changes Everything* (Avery, 2024).

287 *inspired by the research:* Pninit Russo-Netzer and Ofer Israel Atad, "Activating Values Intervention: An Integrative Pathway to Well-Being," *Frontiers in Psychology* 15 (2024), https://www.frontiersin.org/journals/psychology/articles/10.3389/fpsyg.2024.1375237.

NOTES

Epilogue
290 *"This dewdrop world"*: Kobayashi Issa, "This Dewdrop World" in *The Autumn Wind: A Selection from the Poems of Issa*, trans. Lewis Mackenzie (Kodansha International, 1984).

293 *"Fishermen in the cold sea"*: Pablo Neruda, "Keeping Quiet," *The Poetry of Pablo Neruda*, ed. Ilan Stavans (Farrar, Straus, and Giroux, 2003).

Credits and Permissions

All scripture quotations, unless otherwise indicated, are taken from the *New Revised Standard Version Bible*, copyright © 1989 National Council of the Churches of Christ in the United States of America. Used by permission. All rights reserved.

Excerpt from *Letters from a Stoic* by Seneca, translated by Robin Campbell, copyright © Robert Alexander Campbell, 1969. Reprinted with permission from Penguin Random House UK.

Excerpt from Mangala Sutta, *The Suttanipāta*, translated by Bhikkhu Bodhi, copyright © Bhikkhi Bodhi, 2017. Reprinted with permission from Wisdom Publications.

Excerpt from "Zazen on Ching-t'ing Mountain" by Li Po in *Crossing the Yellow River: Three Hundred Poems from the Chinese*, translated by Sam Hamill, copyright © Sam Hill, 2000. Reprinted with permission from Tiger Bark Press.

Excerpt from "Myths" by Anthony De Mello, in *One Minute Wisdom*, copyright © Anthony De Mello, 1985. Reprinted with permission from the DeMello Spirituality Center.

Excerpt from *Standing at the Edge* by Joan Halifax, copyright © Joan Halifax, 2018. Reprinted with permission of Flatiron Books.

Excerpt from "Venus in the Western Sky" by Joseph Goldstein, unpublished. Reprinted with permission from the author.

"Some of us suffer" quote by Swami Beyondananda reprinted with his permission.

Excerpts from *The Art of Forgiveness, Lovingkindness, and Peace*, edited by Jack Kornfield, copyright © Jack Kornfield, 2002. Reprinted with permission from the author.

About the Author

Margaret Cullen is an author, a licensed psychotherapist, and a pioneer in bringing contemplative practices into mainstream settings, codeveloping the Compassion Cultivation Training with Thupten Jinpa at the Stanford University School of Medicine and the Mindfulness-Based Attention Training for military spouses with Amishi Jha at the University of Miami, and founding Compassion Corps, a program that brings compassion programs to underserved populations around the world. She is coauthor of *The Mindfulness-Based Emotional Balance Workbook* (with Gonzalo Brito Pons) and serves as a Mind & Life Institute Fellow and advisory board member of the Global Compassion Coalition. She has been a meditation practitioner for more than forty years.